COLOUR IN DYEHOUSE EFFLUENT

Colour in Dyehouse Effluent

Edited by Peter Cooper

Technical executive, Courtaulds Textiles, Nottingham

1995

Society of Dyers and Colourists

Published by the Society of Dyers and Colourists, PO Box 244, Perkin House, 82 Grattan Road, Bradford, West Yorkshire BD1 2JB, England, on behalf of the Dyers' Company Publications Trust.

This book was produced under the auspices of the Dyers' Company Publications Trust. The Trust was instituted by the Worshipful Company of Dyers of the City of London in 1971 to encourage the publication of textbooks and other aids to learning in the science and technology of colour and coloration and related fields. The Society of Dyers and Colourists acts as trustee to the fund, its Textbooks Committee being the Trust's technical subcommittee.

Typeset by the Society of Dyers and Colourists and printed by The Alden Press, Oxford.

ISBN 0 901956 69 4

Contents

Contributors

Philip C Blowes
Currently director, Arcasorb Technology Ltd, Rickmansworth, Hertfordshire

John H Churchley
Principal process engineer, Severn Trent Water Ltd, Coventry, West Midlands

Keith R F Cockett
Project director, Crosfield Ltd, Warrington, Cheshire

Peter Cooper
Technical executive, Courtaulds Textiles, Nottingham

Clifford Crossley
Applications engineer, PCI Membrane Systems Ltd, Whitchurch, Hampshire

John R Easton
Safety, health and environment advisor, Zeneca Colours, Manchester

John Harrison
Director, Knitting Industries' Federation Ltd, Leicester

Barry G Hazel
Director, The Textile Finishers' Association, Manchester

Peter J Hoyle
Technical representative, Allied Colloids Ltd, Bradford, West Yorkshire

Brian W Iddles
Technical director, Techspan Associates Ltd, Batley, West Yorkshire

Anne Jacques
Project manager, Archaeus Technology Group, Leatherhead, Surrey

David Jones
Engineering manager, Archaeus Technology Group, Leatherhead, Surrey

David M Lewis
Professor of colour chemistry and dyeing, University of Leeds, Leeds, West Yorkshire

Eric J Newton

Managing director, John Heathcoat & Co. Ltd, Tiverton, Devon

John Scotney

Technical director, Stevensons Fashion Dyers, Ambergate, Derbyshire

Timothy G Southern

Proprietor, Environmental Catalysts 2000, Reading, Berkshire

Brian D Waters

Area manager Lower Trent, National Rivers Authority, West Bridgford, Nottinghamshire

Maurice Webb

New opportunities manager, Crosfield Group, Warrington, Cheshire

Peter H Weston

Managing director, Westover Industries Ltd, Great Glen, Leicestershire

Introduction

Peter Cooper

The environmental issues associated with residual colour in treated textile effluents are not new. Indeed, for the last decade or so a number of direct dischargers, both sewage treatment works and commercial textile operations, have had to perform to colour requirements placed on the treated textile effluent discharge, which have been progressively and perceptively tightened. What is perhaps new in the past three to four years in the UK is the higher focus placed by the regulator on public complaints about coloured watercourses and the consequent debate about how control should be exercised to improve the situation and reduce the colour problem to an acceptable level. This debate has widened the traditional discussion/negotiation base from regulator and discharger to regulator, sewerage undertaker and sewer discharger; moreover, because of technical and operational concerns the debate base has widened further to include the dye supplier, trade associations, control equipment suppliers and government departments.

The technical and operational concerns which have widened the debate have included:
- improving the fixation of certain dyes on specific substrates
- chemical structure change of certain dyes to give more biodegradable/bioaccumulated alternatives
- identifying the most cost-effective point in the discharge chain to remove pollution
- the availability of large-scale colour removal processes
- the effectiveness of technical removal techniques to remove various admixtures of dyes
- the 'best practicable environmental option' (BPEO) argument, which suggests that some techniques produce more significant environmental impact (sludge production, landfill options, toxic breakdown products) than that of colour alone
- cost benefits that colour and pollutant removal achieve, particularly with water recycle

- the space implications of new installations, and hence the asset implications of end-of-pipe treatment for some businesses in restricted locations
- the sector infrastructure impact, particularly on small independent businesses, of significant control investment.

The list is by no means exhaustive. It does, however, indicate why there are so many 'players' in the debate and in the search for effective, viable treatments at sewage treatment works or on site, and why the textile operator may feel that there is a need for a clearer picture of the facts of the matter and what the options are for the future.

This publication seeks to put the positions of all the players into perspective and to clarify the current position. It aims to help the textile operator to decide on options available to plan forward strategy that will ensure compliance with the regulators' requirements on a progressive basis. To achieve this objective this publication is divided into parts, each of which consists of chapters in which all the various views, opinions, positions and solutions are explained and developed. It should be remembered that this can only represent a snapshot of a rapidly changing scene in terms of regulations, control technology and cost.

The first chapter in Part 1 presents the dye maker's view, on behalf of the Chemical Industries Association; it discusses the main causes of the problem by class, substantivity and fixation, the biodegradation and/or bioaccumulation and general behaviour of the various dye classes in effluent treatment and structure modification possibilities to cure the problem.

In the second chapter the regulator, the National Rivers Authority (NRA), identifies the need for the control of colour and covers the perceived effect of colour on ecosystems. The colour-measurement techniques employed to quantify the problem and the translation of this into limits are also covered.

The most modern colour removal technology installed at a sewage treatment works in the Severn Trent Water operation in central England is reviewed in Chapter 3. The controls introduced, the effectiveness of the system and the implications of this experience on research and development to control the problem at sewage treatment works are discussed.

Chapter 4 concentrates on the industry view, detailing the 'political' attempt, by the association concerned about the gap between regulatory requirement and technical feasibility, to ensure the industry had a sensible timescale in which to act.

Part 2 begins with a chapter from the Textile Finishers' Association (TFA) that reviews the results of the DEMOS project on waste minimisation sponsored by the UK Department of Trade and Industry. Although originally targeted at identifying the benefits of reduced waste in textile processing, part of the project was focused on the capability and cost implications of the various colour removal techniques available during the timescale of the project. The conclusions are interesting.

A critical review of literature references to colour removal techniques and their relevance to industrial installation and operation is included in Chapter 6. In Chapters 7 and 8, detailed explanations are given of how some of these possible techniques have been developed by two textile-processing businesses with direct discharge to watercourse to achieve the colour standards applied to their particular discharge.

The chapters in Part 3 discuss a selection of currently available technologies in the UK. Reading Chapter 6 may well leave the reader with the impression that there are several techniques to choose from. At the time of writing (late 1994), however, few of these had been proven at semibulk or pilot plant stage. Invitations to contribute to this part of the book were therefore extended to those organisations who had initially shown some success, in the UK, at semibulk or pilot plant scale or larger to single firms or groups of businesses faced with the issue of on-site colour removal. The rationale behind this was that, given the timescales in operation for a solution to the problem, these systems had a much higher likelihood of being available for installation and commission that those that were at the small- or laboratory-scale stage. It is hoped that these contributions give the reader a much clearer idea of the techniques that could be used to overcome the problem, the BPEO implications of any particular technique, the capital and running cost implications to the business and the relevance of each technique to each particular process and dye type.

The writers of Chapters 9 to 14 were asked to cover details of technology type, operational flow diagrams, physicochemical principles involved, results obtained, general cost parameters and opportunities and payback possibilities. To ensure that the general cost parameters in each submission are comparable is extremely difficult; the reader is advised to examine the capital and running costs of each submission in detail, to ensure that comparisons between systems or techniques are based on the same set of parameters. Parameters which are considered key and should be included and directly compared are:

— equipment sized to the same flow rate per operating period

— running costs based on the same operating period

— the level of balancing capability

— sludge disposal, transportation, licence costs

— design, installation, civil and commissioning costs

— depreciation costs

— utility and energy costs

— return on investment based on all capital items.

With this in mind, the reader can compare the various options in terms of both capital and running costs using, perhaps, high, medium and low categorisations. The paper on DEMOS by the TFA attempts to do this equitably, to a degree, but the conclusions of that project are only as good as the information supplied.

The reader is also advised to consider fully the BPEO considerations of the various techniques. There are certainly likely to be different BPEOs for different geographical areas because of issues like dilution and amenity requirement of the receiving water; on the other hand, sludge disposal, incineration, breakdown products, integrated pollution control and treatment chemical residuals may all be more tightly controlled in the near future and these factors could affect final decisions on technical, environmental and cost grounds. Techniques which offer re-use of water help viability and payback, but readers should ensure that the level and area of re-use is acceptable.

A summary or conclusion has not been added to the publication as such a conclusion would be inadvisable at this fluid stage in the cycle of events. Some things can be identified, however. The problem has to be solved to the satisfaction of all the 'players'; otherwise the impact – on the textile sector in particular – could be detrimental. Thus the regulatory, technical and cost implications on treatment on-site and at sewage treatment works must be very carefully considered before irrevocable decisions are made.

There are both threats and opportunities. The threats are tighter regulatory control, increased costs (whether from extra sewage treatment works or on-site treatments) and the consequences of those costs on profitability. The opportunities, certainly from some of the emerging technologies, are cost-beneficial from re-use, recovery and optimisation.

Initial end-of-pipe solutions will differ according to the very varied admixtures of dyes in effluent discharged from the very varied textile processors. The best removal may be obtained from a combination of technologies rather than from one single-stage process, even if those separate steps include chemical dosing to change the character of the effluent.

Control of colour in effluent is likely to be of increasing importance nationally and internationally. New installations will receive colour consents prior to start-up, and therefore the infrastructure of these new installations will need to be designed differently from the conventional designs of today. Segregated waste water from dyeing, holding and balancing capacities and drain construction and layout may be appropriate examples. This may well put a different cost base to these new installations. Process optimisation and waste minimisation will therefore require improved focus.

Control of colour may instigate some lateral thinking. Chapter 15 gives an example of a possible approach.

Part One

The problem of colour

Chapter 1

The dye maker's view, *John R Easton, Zeneca Specialties, Manchester*

This article highlights the diversity of the chemistry used in textile dyes and reviews the studies carried out to assess the impact of dyes released into the aquatic environment. The main problem is one of visual pollution, particularly arising from reactive dyes due to their relatively lower levels of fixation compared to other dye classes and their poor removal at sewage treatment works. The dye manufacturers have a role to play in reducing environmental impact of the dyeing process by developing new products, developing application processes to maximise right-first-time production, and evaluating effluent decolorisation techniques.

Chapter 2

The regulator's view, *Brian D Waters, National Rivers Authority, Severn-Trent Region, Nottingham*

Colour in watercourses rising from dyeing and related activities is an increasing problem for the National Rivers Authority. Changes in dyeing practice may be partially to blame but there are increasing complaints from the public on the issue. Toxicity is not usually a problem, and it is primarily the aesthetic effects that need to be tackled. This makes the setting of appropriate standards difficult, as the perception may be affected by other features of a watercourse.

Chapter 3

The water company's view, *John H Churchley, Severn Trent Water Ltd, Coventry*

Colour removal has been successfully achieved at Leek sewage treatment works using ozone generated from air. The high capital cost of ozone treatment means that its use is confined to those sewages with a high proportion of dye waste. Chemical coagulation/flocculation has been unsuccessful at certain sites but appeared promising in a small-scale trial at Wanlip (Leicester) sewage treatment works. Colour removal at the dyehouse is the option preferred by water companies, but decolorisation at sewage works may be practised as an interim measure.

Chapter 4

The trade association's view, *John Harrison, Knitting Industries' Federation Ltd, Leicester*

The trade association's attempts to bridge the gap between regulatory requirements and technical feasibility are described, including the 'political' approach used to ensure that the appropriate message was understood at both legislatory and regulatory levels.

CHAPTER ONE

The dye maker's view

John R Easton

1 Introduction

Since the foundation of the synthetic dye industry with Perkin's synthesis of mauveine in 1856, generations of chemists have applied their minds to the challenge of designing dyes for an ever-widening range of fibre substrates and application methods. The large number of dyes in use today bears witness to their creativity and innovation in successfully meeting this challenge and satisfying the dyer's demands for simple, reproducible application processes, and the consumer's need for quality products at a reasonable price.

Dyes cover a broad range of chemical types, and the system of classification which has gained international acceptance is that used in the *Colour Index*. It classifies colorants in two different ways, assigning to each firstly a CI Generic Name, based on its application characteristics (Table 1), and secondly a CI Constitution Number based on its chemical structure (where known) (Table 2) [1].

Table 1 *Colour Index* Generic Names for major dyestuff groups	
Acid	Pigment
Basic	Reactive
Direct	Solvent
Disperse	Sulphur
Mordant	Vat

Shore has pointed out that if colorant precursors (such as azoic components and oxidation bases) are excluded, as well as the sulphur dyes of indeterminate constitution, almost two-thirds of all the organic colorants listed in the *Colour Index* are azo dyes, one-sixth of them being metal complexes [2]. The next largest class is that of anthraquinone dyes (15% of the total) followed by triarylmethanes (3%) and phthalocyanines (2%). Table 3 shows the percentage distribution of each chemical class between major application ranges [2].

Of the 3000 or so distinct chemical entities in the *Colour Index*, only about 30 are used at a rate in excess of 1000 tonnes per annum, and 90% of the products are used at the level of 100 tonnes per annum or less [3].

Table 2 *Colour Index* classification of chemical constitutions

Chemical class	CI Constitution Numbers	Chemical class	CI Constitution Numbers
Nitroso	10000–10299	Indamine	49400–49699
Nitro	10300–10099	Indophenol	49700–49999
Monoazo	11000–19999	Azine	50000–50999
Disazo	20000–29999	Oxazine	51000–51999
Triazo	30000–24999	Thiazine	52000–52999
Polyazo	35000–36999	Sulphur	53000–54999
Azoic	37000–39999	Lactone	55000–55999
Stilbene	40000–40799	Aminoketone	56000–56999
Carotenoid	40800–40999	Hydroxyketone	57000–57999
Diphenylmethane	41000–41999	Anthraquinone	58000–72999
Triarylmethane	42000–44999	Indigoid	73000–73999
Xanthene	45000–44999	Phthalocyanine	74000–74999
Acridine	46000–46999	Natural	75000–75999
Quinoline	47000–47999	Oxidation base	76000–76999
Methine	48000–48999	Inorganic pigment	77000–77999
Thiazole	49000–49399		

Table 3 Distribution of each chemical class between major application ranges [2]

Chemical class	Distribution between application ranges/%								
	Acid	Basic	Direct	Disperse	Mordant	Pigment	Reactive	Solvent	Vat
Unmetallised azo	20	5	30	12	12	6	10	5	
Metal-complex azo	65		10				12	13	
Thiazole		5	95						
Stilbene			98						
Anthraquinone	15	2		25	3	4	6	9	36
Indigoid	2					17			81
Quinophthalene	30	20		40			10		
Aminoketone	11			40	8		3	8	30
Phthalocyanine	14	4	8		4	9	43	15	3
Formazan	70						30		
Methine		71		23		1		5	
Nitro, nitroso	31	2		48	2	5		12	
Triarylmethane	35	22	1	1	24	5		12	
Xanthene	33	16			9	2	2	38	
Acridine		92		4				4	
Azine	39	39				3		19	
Oxazine		22	17	2	40	9	10		
Thiazine		55			10			10	25

The global textile coloration industry is therefore characterised by a very large number of geographically dispersed dyehouses of small to medium size, using a very wide range of dyes of diverse chemical constitution.

2   The release of dyes into the aquatic environment

The principal route by which dyes enter the environment is via waste water from batch processes in both the dye-manufacturing and the dye-consuming industries. In 1978 it was estimated that of the 450 000 tonnes of dye produced worldwide, some 9000 tonnes (2%) were discharged in aqueous effluents from manufacturing operations and 40 000 tonnes (9%) from the coloration sector [4].

In the UK the vast bulk of such releases are to the sewerage system; dyehouse effluents are treated along with domestic and other industrial effluents in sewage treatment works, before being discharged to surface waters. Until the water industry was privatised in 1989 sewage treatment works were operated by the regional water authorities, but they are now run by private water companies. The policing of discharges to surface waters is now carried out by the National Rivers Authority (NRA).

For textile dyeing operations Brown has indicated that the predicted environmental concentration (PEC) of a dye in a receiving water may be calculated from the following factors [4]:
— daily usage of the dye
— degree of fixation on the substrate
— degree of removal during any effluent treatment process
— dilution factor in the receiving water.

A combination of average values produced a figure of 1 mg l^{-1} for a single dye in a typical river. This is clearly very low; it is an average annual value, however, and since textile dyeing operations are characteristically batch processes, there are occasions when dye concentrations may be much higher than this. Hobbs has described such scenarios (Table 4) in her paper based on a survey of the dye industry in the UK [5]. More recently Pierce has noted the peak loads of reactive dye colour discharged from a typical cotton dyehouse [6].

3   Impact of dyes on the environment

Dyes are required to show a high degree of chemical and photolytic stability in order to fulfil the fastness requirements of both retailers and consumers. One consequence of this stability is that they are not readily degraded under the aerobic conditions prevailing in the biological

11

Table 4 Examples of calculated estimates of dye concentration in river water as a result of dye release by the textile industry[a] [5]

Scenario[b]	Concentration of dye in receiving river/mg l^{-1}	No of days each year
1 – average	5.3	50
1 – worst case	1555	2
2 – average	1.2	25
2 – worst case	364	2

a Assuming no adsorption of dye on the sludge at the sewage treatment works
b 1 – Batchwise dyeing of cotton with reactive dyes;
   2 – Batchwise dyeing of wool yarn with acid dyes

treatment plant at a sewage treatment works. So unless colour is removed by chemical or physical means, either at the dyehouse or by tertiary treatment at the sewage treatment works, it may well pass through to the receiving water and lead to the kind of public complaints which in the UK have prompted the NRA action on colour standards in rivers. Concentrations of dye as low as 1 mg l^{-1} can give rise to public complaint. 'Unnatural' colours such as red and purple usually cause most concern; the public seems more ready to accept blue, green or brown rivers.

It is generally well accepted that colour in waterways is an aesthetic problem rather than a ecotoxic hazard. The Ecological and Toxicological Association of Dyes and Organic Pigment Manufacturers (ETAD) has, over the last 20 years, been pre-eminent in promoting an understanding of the human health and environmental impacts of colorants and contributing to the body of knowledge in these areas. ETAD, in a review of the data available on fish toxicity, found that 98% of dyes have an LC_{50} value greater than 1 mg l^{-1}, and that 59% have an LC_{50} value greater than 100 mg l^{-1} (Table 5). Clarke and Anliker have summarised the findings [7].

Table 5 Fish toxicity levels of dyes

LC_{50} value /mg l^{-1}	Proportion of dyes/%
<1	2
1–10	10
10–100	27
100–500	31
>500	28

ETAD has also investigated the effect of dyes on waste water bacteria to determine whether dye reaching a sewage treatment works could adversely affect the operation of the biological process [8]. Of over 200 dyes tested in the ETAD study, only 18 showed significant inhibition of the respiration rate of the biomass, and these were all basic dyes.

No long-term tests have been undertaken to study possible sublethal, chronic effects of dyes on fish, but the partition coefficient of a dye in n-octanol/water (P_{ow}) may be used as a measure of bioaccumulation potential. If the P_{ow} value is less than 1000 (i.e. $\log P_{ow} < 3$) then it can be confidently predicted that the *bioconcentration factor* (BCF) is less than 100 and the dye is unlikely to bioaccumulate. (Bioconcentration is the uptake and retention of a chemical by a living organism from its immediate environment.[1]) The BCF is normally measured in fish and is the ratio at equilibrium (concentration of the chemical found in the fish):(concentration of the chemical in the water).

No bioaccumulation is expected for dyes with a water solubility greater than 2000 mg l^{-1}. Even water-insoluble disperse dyes with P_{ow} values greater than 1000 have been shown not to bioaccumulate. The various reasons for this have been reviewed [9,10], as has the relationship of bioaccumulation to hydrophobicity and steric factors [11].

There are relatively few reports of measured dye levels in rivers [12]. ETAD has recognised the importance of obtaining reliable field data and has supported the development of appropriate analytical techniques. In a recent project four member companies have each studied a single dye and monitored releases from manufacturing plants and textile dyehouses [13].

But despite the tremendous advances in analytical methods over the last few years, the difficulties associated with the identification of individual *unknown* dyes in a coloured effluent or watercourse remain considerable.

It is open to question whether the development of such techniques should be a high priority when it is clear from the studies cited above that for the vast majority of dyes toxic effects in the aquatic environment are not the issue.

4 Dye/substrate effects

One of the major factors determining the release of a dye to the environment is its degree of fixation on the substrate. There is a wide divergence of 'average' or 'typical' values amongst the many possible dye/substrate combinations, since fixation is clearly dependent on individual circumstances and governed by several factors including depth of shade,

---

[1] Biomagnification is the term applied to the uptake and retention of a chemical by a living organism from its food. Hence bioaccumulation is a combination of bioconcentration and biomagnification. It is well accepted that for aquatic organisms the major pathway for bioaccumulation is from the surrounding water, i.e. bioconcentration.

Table 6 Estimated degree of fixation for different dye/fibre combinations

Dye class	Fibre	Degree of fixation/%	Loss to effluent/%
Acid	Polyamide	80–95	5–20
Basic	Acrylic	95–100	0–5
Direct	Cellulose	70–95	5–30
Disperse	Polyester	90–100	0–10
Metal-complex	Wool	90–98	2–10
Reactive	Cellulose	50–90	10–50
Sulphur	Cellulose	60–90	10–40
Vat	Cellulose	80–95	5–20

application method and liquor ratio. It is therefore impossible to give a definitive set of figures. Table 6 represents the author's own summary of the data available.

The problem of coloured effluent has become identified particularly with the dyeing of cellulose fibres (notably cotton, which accounts for almost 50% of the total fibre consumed by the textile industry worldwide), and in particular with the use of reactive dyes. It can be seen from Table 6 that as much as 50% of reactive dye used may be lost to the effluent.

For all dye/fibre combinations the degree of fixation decreases with increasing depth of shade. The demand for heavy red, navy and black shades coming down the supply chain from consumers, designers and retailers can thus have a significant effect on the amount of dye that is released to the effluent in the commission dyehouse. It is therefore important that those involved in the demand side should be aware of the consequences of their fabric and colour choices on the production operation and its associated environmental impacts.

This does not, however, absolve the dye manufacturers of their responsibility to introduce new dyes, particularly reactive dyes, showing improved levels of exhaustion and fixation over those currently available to the industry.

5 Dye adsorption in biological treatment plant

As already mentioned, dyes are not readily biodegraded in aerobic sewage treatment processes. Some, however, may be adsorbed intact by the sludge at a sewage treatment works and hence removed from the aquatic compartment. This sorptive removal of dyes is often referred to as *bioelimination*; it varies widely from dye to dye, with no clear structural correlation to explain their behaviour.

Hitz *et al.* studied bioelimination of various dyes from different classes and reached the following conclusions [14]:

(a) acid dyes – high solubility leads to low adsorption, which appears to depend on the degree of sulphonation

(b) reactive dyes – very low degree of adsorption, apparently unaffected by the degree of sulphonation or ease of hydrolysis

(c) basic dyes – typically high levels of adsorption

(d) disperse dyes – adsorption in the high-to-medium range

(e) direct dyes – high degree of adsorption, apparently unrelated to the number of sulphonic acid groups.

Pagga and Brown reported studies on 87 dyes and found that 62% showed significant levels of colour removal by adsorption in a modified Zahn-Wellens test [15].

Shaul *et al.* reported on investigations into the partitioning of water-soluble azo dyes in the activated sludge process [16]. A total of 18 dyes were tested and categorised according to their behaviour in the tests (Table 7). Again it was concluded that the high degree of sulphonation of the azo dyes in Group 1 enhanced their water solubility and limited their ability to adsorb on the biomass. Although dyes in Group 2 were also highly sulphonated, their greater molecular size was thought to account for their greater degree of adsorption.

Table 7 Fate of water-soluble azo dyes in the activated sludge process

Group 1[a]	Group 2[b]	Group 3[c]
CI Acid Yellow 17	CI Acid Red 151	CI Acid Orange 7
CI Acid Yellow 23	CI Acid Blue 113	CI Acid Orange 8
CI Acid Yellow 49	CI Direct Yellow 28	CI Acid Red 88
CI Acid Yellow 151	CI Direct Violet 9	
CI Acid Orange 10		
CI Acid Red		
CI Acid Red 14		
CI Acid Red 18		
CI Acid Red 337		
CI Acid Black 1		
CI Direct Yellow 4		

a Dyes passing through essentially unaffected
b Dyes eliminated by adsorption
c Dyes showing evidence of biodegradation

Information on the ecotoxicology of dyes is given in Section 12 of the recently introduced standard format Safety Data Sheet prescribed in Europe [17]. An analysis of the bioelimination data available on Zeneca Colours' Procion reactive dyes leads to the following conclusions:

(a) monoazo dyes are particularly poorly adsorbed

(b) disazo dyes are generally much better adsorbed than monoazos

(c) anthraquinones, triphendioxazines and phthalocyanines are all adsorbed to a much higher degree than monoazos

(d) there does not appear to be a direct correlation between bioelimination and either number of sulphonic acid groups or molecular mass.

The chromophore distribution in reactive dyes (Table 8) indicates that the great majority of unmetallised azo dyes are yellows, oranges and reds. In contrast the blues, greens, blacks and browns contain a much higher proportion of metal-complex azo, anthraquinone, triphendioxazine or copper phthalocyanine chromophores. It is possible that preferential adsorption of these latter chromophore types from a coloured dyehouse effluent during biological treatment leads to the characteristic straw or pink tones often observed in the sewage treatment works outfall.

Further studies are required on water-soluble acid and reactive dyes in order to gain a better understanding of the structural features that determine their adsorption behaviour in activated sludge processes.

Table 8 Distribution of chemical classes in reactive dye hue sectors [2]

Chemical class	Distribution in hue sector/%								Proportion of all reactive dyes/%
	Yellow	Orange	Red	Violet	Blue	Green	Brown	Black	
Unmetallised azo	97	90	90	63	20	16	57	42	66
Metal-complex azo	2	10	9	32	17	5	43	55	15
Anthraquinone				5	34	37		3	10
Phthalocyanine					27	42			8
Miscellanous	1		1		2				1

6 Anaerobic degradation

Although in general dyes show very little evidence of biodegradation under aerobic conditions, many azo dyes can be readily degraded under the anaerobic conditions found in sewage sludge digesters or sediments in surface water (for example, river muds). Work has been carried out by ETAD to assess the primary degradability of 22 water-soluble dyes under anaerobic conditions [18]. The results showed that, with the single exception of the anthraquinone acid dye CI Acid Blue 80, all the dyes tested showed a substantial degree of colour removal. The authors concluded that the breakdown of dyes in the environment is likely to be initiated under anaerobic conditions.

There is concern that the aromatic amines, which are formed as metabolites by reductive fission of the azo bond under anaerobic conditions, could pose a more serious toxic hazard in the aquatic environment than the intact dye molecules. ETAD therefore went on to study the aerobic degradation of simple aromatic amines and found that they were unlikely to be persistent. The studies were also extended to investigate the fate of a wider range of aromatic amines including sulphonated species [19,20].

The use of sequential anaerobic-aerobic systems for the treatment of textile effluents has been described in the literature [21], and is in operation at a major new dyehouse in Hong Kong [22]. Research into the detailed mechanisms of anaerobic decolorisation of dyes is continuing in order to support the development of this technology for the treatment of dyehouse effluents [23].

7 The role of the dye manufacturer

In recognising that visual impairment of surface waters by unnatural coloration is a matter of genuine public concern, what contribution can the dye manufacturer make to the resolution of the problem? There are perhaps three major areas:
- research and development of new dyes
- application and process development
- advice on waste minimisation and effluent treatment.

7.1 Research and development

The European dye manufacturing industry has an outstanding record of innovation stretching back many years. One measure of this is the fact that of all new substances notified in Europe under the so-called Sixth Amendment since 1983, around 20% are colorants.

Dyes showing a high level of exhaustion and fixation of the dye on the fibre have been, and will continue to be, prime targets of research and development activity. Notable improvements have been achieved over the last twenty years or so in the field of reactive dyes by the development of dyes containing more than one reactive group [24].

If we assume a fixation efficiency of 70% for an individual reactive group and a typical dye

17

exhaustion of 85%, then the overall fixation of the monofunctional dye is 59.5%; in other words, 40.5% of the dye is lost to the effluent.

In the case of a bifunctional dye containing two reactive groups of similar reactivity, the further possibility for fixation introduced by the second reactive group raises the fixation efficiency to 91% [70% + (30% × 70%)]. Assuming a degree of exhaustion equal to that of the monofunctional dye, then the overall fixation of the bifunctional dye is 77% and only 23% of the dye is lost to the effluent. That is, addition of the second reactive group produces a reduction of 40% in the amount of colour lost to the effluent.

Bifunctional dyes containing two identical reactive systems are referred to as *homobifunctional*. They are exemplified by the Procion H-E dyes developed by ICI (now Zeneca) in the late 1970s and by certain bisvinylsulphone dyes, the most notable example being CI Reactive Black 5 originally developed by Hoechst. The first range of *hetero-bifunctional* dyes (dyes containing two different reactive systems) was the Sumifix Supra range introduced by Sumitomo in 1980 [25]; other manufacturers have since developed the concept, such as Ciba with Cibacron C dyes [26]. But a major step-change in technology would be required to deliver reactive dyes with greater than 95% overall fixation across the full shade gamut under all application conditions. An invention of this kind is impossible to predict but is unlikely to be achieved before the end of the century.

A further factor to be borne in mind is that before any new substance can be marketed in Europe a considerable amount of time and money has to be spent on toxicological and ecotoxicological testing. The notification requirements have recently been modified in the UK with the introduction of the Notification of New Substance Regulations 1994 [27], which implement the Seventh Amendment of the EC Dangerous Substances Directive (EEC/92/32) [28]. An outline of the data required to be provided by the notifier is given in Table 9. Clearly, the compilation of the information dossier required for the notification of a new substance in Europe is both expensive and time-consuming, a situation which militates against the rapid replacement of whole product ranges.

7.2 Application process development

The search for dynamic response and improved productivity has served to focus the attention of the coloration industry on right-first-time (RFT) production techniques [29–32]. By

---

Table 9 Data on new substances to be provided by the notifier

---

1 *Identity of the substance*
– Name(s), trade name(s), Chemical Abstracts Service (CAS) number, molecular and structural formula
– Composition of the substance, degree of purity, nature of impurities, spectral data (UV, IR, NMR, MS)
– Methods of detection and determination

2 *Information on the substance*
– Production process, exposure estimates related to production
– Proposed uses, types of use (function of the substance and desired effects), exposure estimates related to use, fields of application, waste quantities and composition of wastes resulting from proposed uses
– Form under which the substance is marketed: substance, preparation, products
– Concentration of the substance in commercially available preparations
– Estimated production and/or imports for each of the anticipated uses or fields of application
– Overall production and/or imports in tonnes per year for the first calendar year and subsequent years
– Recommended methods and precautions concerning handling, storage, transport, fire and other dangers
– Emergency measures to be taken in the case of accidental spillage or injury to persons, e.g. poisoning

3 *Physico-chemical properties of the substance*
– Physical form of the substance: melting point, boiling point, relative density, vapour pressure, surface tension, water solubility, partition coefficient (*n*-octanol/water), flash point, flammability, explosive properties, self-ignition temperature, oxidising properties, granulometry, particle size distribution

4 *Toxicological studies*
– Acute toxicity: oral, inhalation, dermal
– Skin irritation, eye irritation
– Skin sensitisation
– Repeated dose (28 days) toxicity
– Mutagenicity
– Toxicity to reproduction
– Toxicokinetic behaviour

5 *Ecotoxicological studies*
– Acute toxicity to fish
– Acute toxicity to daphnia
– Growth inhibition test on algae
– Bacteriological inhibition
– Degradation: ready biodegradability, abiotic degradation (hydrolysis as a function of pH), BOD, COD, BOD/COD ratio

---

definition, a high level of RFT minimises waste and can make a contribution to reduced colour loads in the effluent [33]. Much has been achieved by collaborative efforts between dye and machinery manufacturers in Europe in designing products and processes offering greater control of the dyeing or printing operation. A wealth of technical information and advice on the optimisation of processes and minimising of dye wastage is available to the textile industry from the major dye suppliers.

For example, in the UK the Knitting Industries' Federation has recently produced a *Water and effluent management manual*, with financial assistance from the Department of Trade and Industry and with the co-operation of several major dye manufacturers [34]. As well as incorporating 'best practice' advice from individual manufacturers for their particular product range, the *Manual* also covers general principles of good housekeeping and safe handling and disposal, drawing on information made available by ETAD and the Chemical Industries Association (CIA) Dyes Sector Group.

It is noteworthy that in other areas of emerging technology, such as ink-jet printing of textiles or dyeing from supercritical carbon dioxide, collaboration along the supply chain between the dye suppliers, the machinery manufacturers, and the textile dyer or printer has been a key feature of the development process.

7.3 Advice on effluent decolorisation

A range of technologies for decolorisation has been investigated by the major dye suppliers, partly with a view to ensuring compliance of the waste water discharges from their own manufacturing sites but also with the intention of assisting customers who are under pressure to reduce colour levels in their effluents [3,35–40]. There is continuing co-operation with academic institutions, water companies, waste water treatment plant suppliers and others in seeking to develop technical solutions for the treatment of coloured effluents [40].

8 Conclusion

There is no quick fix to the complex problem of colour in dyehouse effluent. Only by means of a collaborative partnership between dye suppliers, machinery manufacturers, water treatment companies and the textile industry itself, including designers and retailers, will satisfactory solutions be found. For their part, the dye makers are already committed to such partnerships under the guiding principles of the chemical industry's Responsible Care and Product Stewardship programmes.

9 References

1.   *Colour Index*, 3rd Edn, Vol. 4 (1971), Vol. 9 (1992) (Bradford: SDC).
2.   *Colorants and auxiliaries*, Vol. 1, Ed. J Shore (Bradford: SDC, 1990).

3. I G Laing, *Rev. Prog. Coloration*, **21** (1991) 56.

4. D Brown, *Ecotox. Environ. Safety*, **13** (1987) 139.

5. S J Hobbs, *J.S.D.C.*, 105 (1989) 355.

6. J H Pierce, *J.S.D.C.*, 110 (1994) 131.

7. E A Clarke and R Anliker, *Rev. Prog. Coloration*, **14** (1984) 84.

8. D Brown, H R Hitz and L Schaefer, *Chemosphere*, **10** (1981) 263.

9. R Anliker, E A Clarke and P Moser, *Chemosphere*, **10** (1981) 263.

10. R Anliker and P Moser, *Ecotox. Environ. Safety*, **13** (1987) 43.

11. R Anliker, P Moser and D Poppinger, *Chemosphere*, **17** (1988) 1631.

12. D Brown and R Anliker, Royal Soc. Chemistry symposium, *Risk assessment of chemicals in the environment*, Ed. M L Richardson (London: 1988), 398.

13. ETAD 20th Annual Report, Basle (1994) 22.

14. H R Hitz, W Huber and R H Reed, *J.S.D.C.*, **94** (1978) 71.

15. U Pagga and D Brown, *Chemosphere*, **15** (1986) 479.

16. G M Shaul, C R Dempsey and K A Dostal, EPA Project Summary EPA/600/S2-88/030 (Cincinnati: EPA, 1988).

17. Material Safety Data Sheet Directive (91/155/EEC), OJEC L76/55 (1991).

18. D Brown and P Laboureur, *Chemosphere*, **12** (1983) 397.

19. D Brown and P Laboureur, *Chemosphere*, **12** (1983) 405.

20. D Brown and B Hamburger, *Chemosphere*, **16** (1987) 1539.

21. Yand Xikun, *Text. Asia*, (Oct 1990) 178.

22. *Text. Asia*, (Sept 1993) 18.

23. C Carliell, MScEng thesis, University of Natal (1993).

24. A H M Renfrew and J A Taylor, *Rev. Prog. Coloration*, **20** (1990) 1.

25. S Fujioka and S Abeta, *Dyes Pigments*, **3** (1982) 281.

26. J P Luttringer and A Tzikas, *Textilveredlung*, **25** (1990) 311.

27. The Notification of New Substance Regulations 1993, SI 1993, No 3050 (London: HMSO).

28. Council directive 92/32/EEC amending for the seventh time directive 67/548/EEC on the approximation of laws, regulations and administrative provisions relating to the classification, packaging and labelling of dangerous substances, OJEC L154 (1992).

29. P S Collishaw, B Glover and M J Bradbury, *J.S.D.C.*, **108** (1992) 13.

30. P S Collishaw, D A S Phillips and M J Bradbury, *J.S.D.C.*, **109** (1993) 285.

31. K Parton, *J.S.D.C.*, **110** (1994) 4.

32. M J Bradbury and J Kent, *J.S.D.C.*, **110** (1994) 173, 222.

33. B Glover and L Hill, *Text. Chem. Col.*, **25** (1993) 15.

34. J K Skelly, Presentation to DTI symposium on 'Colour in products: environmental lifecycle perspectives', Brighouse (June 1994).

35. U Sewekow and W Beckmann, *Textil Praxis*, **46** (1991) 346, 445.

36. A Reife in *Kirk-Othmer encyclopedia of chemical technology*, 4th Edn, Vol. 8 (New York: John Wiley, 1993) 753.

37. D Tegtmeyer, *Melliand Textilber.*, **74** (1993) 148.

38. U Sewekow, *Melliand Textilber.*, **74** (1993) 153.

39. U Sewekow, AATCC Int. Conf., *Book of papers*, Montreal (Oct 1993) 235.

40. D L Woerner, L Farias and W Hunter, *Utilisation of membrane filtration for dyebath reuse and pollution prevention*, AATCC Symposium, Charlotte, NC (Apr 1994).

CHAPTER TWO

The regulator's view

Brian D Waters

1 Introduction

Coloured effluents have probably been produced since dyeing first began. The earliest attempts at dyeing were on a very small scale, however, the dyes were natural products of limited colour intensity and the consequential environmental impact would have been minimal. Over several hundred years the scale of operation, the nature of the dyes and the impact have all changed dramatically. There is increasing public pressure for improvements in river water quality and the National Rivers Authority (NRA) has been pressing for further improvements in the quality of discharges.

2 The National Rivers Authority

The NRA came into existence in September 1989 as the public body responsible for the management of the water environment. It was formed from those parts of the regional water authorities that were responsible for the management of the water environment at the time of the creation and privatisation of the water utility companies. Its responsibilities extend to all 'controlled' waters, that is, rivers and streams, canals, groundwaters and coastal waters.

The duties of the NRA are extensive:
- maintenance of water quality and pollution control
- management of water resources and licensing of abstractions
- construction and maintenance of flood defences
- maintenance, improvement and development of fisheries
- promotion of conservation and recreation in the aquatic environment
- navigation (some catchments only).

These activities interact with each other, and at times their objectives are in conflict. The NRA has to balance their demands, while providing for new requirements, particularly of water supplies and of means of effluent disposal, in a sustainable way. This is being tackled through a catchment management process involving public consultation to develop catchment plans for the next decade or so.

3 The nature of the problem

The presence of colour as a water quality problem was briefly mentioned above. There has been much interest in the Midlands since the 1970s, increasing in importance in recent years [1–3]. The reasons for the rising awareness of the problem are complex, but can best be summed up as an apparent deterioration in quality at a time of increasing environmental expectations.

The apparently greater impact of coloured effluents may be due to one or more of several factors, described below.

3.1 Industrial concentration

The dyeing industry is located in only a few towns. This is largely for historical reasons, the early requirements of the industry being the availability of soft water and a means of effluent disposal. Although these factors are less critical than they were during the Industrial Revolution, the textile industry has continued to grow in the traditional locations, probably due nowadays to economic factors and the availability of a skilled workforce. Whatever the reason, textile processing, including dyeing, is concentrated as shown in Table 1 (this list is not exhaustive). For example, Leek in

Table 1 Major textile-producing areas in the UK

Area	Industry
West Yorkshire	Wool
Southern Scotland	Wool
Lancashire	Cotton
East Midlands	Hosiery and knitwear
Kidderminster	Carpets

Staffordshire has experienced major discharge problems, with dyeing and dye manufacture accounting for about 60% of the effluent discharged to the local sewage treatment works. This in turn discharges to the River Churnet with very low dilution.

3.2 The nature of the dyes

Modern society has become more demanding of the standards of clothes and other fabrics that it buys. The customer wants bright colours that will survive severe treatment; stability to withstand repeated washing is required, together with the ability to meet the greater and greater demands placed on dye wet fastness by today's detergents, particularly by the oxidative systems they incorporate. As the fastness of dyes has increased to meet this demand, they have naturally become less amenable to the aerobic biological processes commonly used in sewage and industrial effluent treatment.

Water-soluble dyes are a particular problem, though finely divided insoluble dyes and pigments can be difficult to settle. The mechanism for the removal of soluble dyes by an aerobic biological treatment may be straightforward biodegradation, but is mainly by adsorption on to the biomass and destruction in subsequent sludge processing. The term *bioelimination* may be used to describe collectively these different mechanisms (different writers use this term in slightly different senses, however). Most classes of water-soluble dyes are removed from effluent streams by bioelimination, but reactive dyes form a notable exception. As a class these dyes are removed to a maximum of approximately 30% and an average of only 10%. They thus pass directly through a biological treatment plant into the receiving river.

The problem with reactive dyes is exacerbated by their relatively low fixation to cellulosic fibres, in general 70–80%, compared with other dye classes and fibres. Colour in rivers is thus likely to present the greatest difficulties in geographical areas with a high concentration of dyehouses dyeing cotton using reactive dyes.

3.3 Public expectations

Contrary to the generally accepted image, river water quality has improved significantly over recent years. The measures of quality have changed over time but broadly comparable trends can be identified since monitoring became established on a systematic basis about thirty years ago. River water quality has been classified into bands since 1958. In 1980 the National Water Council introduced a scheme with four main bands (Table 2) [4]. The bands were defined in terms of the chemical quality of the water using as criteria biological oxygen demand (BOD), dissolved oxygen content and ammonia content.

The progressive improvement in quality can be seen from the changing proportions of river length in each class over time [5,6]. Although the methodology of assessment has changed over time and has not been applied on a strictly consistent basis across the country, within a catchment long-term comparisons can be made with a reasonable degree of confidence. Within the catchment of the River Trent (which includes rivers affected by dyes such as the Churnet, Soar and Erewash), there has been a reduction in the proportions of length in classes 3 and 4 and an increase in the proportions in classes 1 and 2. The distribution is approaching that required to meet the river quality objectives (RQOs) (Table 3).

The improvement in quality has been most marked in the industrial areas, to the extent that fish are now found in rivers that were fishless only a few years ago. This brings to the river bank not only fishermen but also the broader public wishing to walk, picnic, use boats and canoes and observe the increasing wildlife. This public has a better awareness of environmental issues; indeed, these are currently very much to the fore in education. Consequently they have greatly enhanced expectations and will complain about dead fish and other wildlife, and the presence of unusual indications of changes in quality such as odour and colour. If colour is changing over time it is even more likely to be noticed and reported.

Research conducted for the NRA has found that public perception of water quality is very influenced by colour, as purity is associated with rushing water and clarity [7]. Unnatural colour is associated with contamination and, if standards are to be set, public acceptability needs to be taken into account.

The NRA and its predecessor organisations have not identified complaints of colour on a systematic basis. In 1992, a survey highlighted up to 700 complaints a year nationally, affecting over 600 km of river [7]. The highest numbers were in the Severn-Trent region. Over the last few years colour has accounted for about 5% of the total number of complaints

Table 2 National Water Council classification of river water quality

Description	Class
Good	1 (a and b)
Fair	2
Poor	3
Bad	4

Table 3 Condition of Severn-Trent rivers compared with RQO

Year	Class 1	Class 2	Class 3	Class 4
	Length of river in each class/%			
1961	22	8	23	47
1974	23	42	13	22
1979	22	57	13	8
1990	24	51	19	6
RQO	25	68	7	

received. This is less than for other generic sources such as oil and sewage, but these tend to be one-off incidents whereas colour is usually traceable to a limited number of persistent point sources. There have also been specific complaints by abstractors treating water for potable and industrial supplies, as conventional water treatment processes are normally unable to remove dyes.

3.4 Toxicity

There is little evidence that the dyes found in watercourses are toxic to fish and other wildlife at the concentrations likely to be present. The published toxicity data also support this view. It has been suggested that strong colours could reduce light penetration, thus affecting the growth of plants and impacting on invertebrate and other forms of wildlife. The stretches of river where there is poor biology associated with dyes in the river may also be affected by other forms of pollution, including the presence of permethrin from mothproofing. Consequently the allegations of toxicity due to dyes are not proven.

4 River water quality objectives

Historically river water quality was measured against RQOs using the National Water Council Scheme established in 1975 [4]. These, the standards referred to above, are being replaced now by statutory water quality objectives (SWQOs) and a new general quality assessment (GQA) [8]. SWQOs will be a set of use-related standards, incorporating directives from the European Union. They will be set by the Secretaries of State for the Environment and for Wales on advice from the NRA. To date the only scheme approved is The Surface Waters (Ecosystem) (Classification) Regulations 1994, but no SWQOs have yet been set [9].

European directives applying to the protection of fisheries [10], and to minimum standards of treatment for surface waters used for the supply of drinking water [11] and the discharge of dangerous substances [12], are also in force.

The GQA system is similar in many ways to the RQO system in that it uses biological oxygen demand, dissolved oxygen and ammonia as the basis for classification. The main difference is that the original four classes are replaced by six, with a more rigorous definition and implementation of the classification system. It is intended that the chemical classification system be complemented with measures of biological quality, nutrient status and aesthetic quality in due course. The last of these will probably include colour, and the NRA has been investigating how this can best be achieved.

5 Setting colour standards

Most of the standards applied to water are expressed in milligrams per litre (mg l^{-1}) of the determinand, with obvious exceptions such as temperature and pH. Trying to express colour standards in this way is not practicable, however, as the same concentrations of different dyes may give rise to different colour intensities of the same shade or to completely different hues. The research for the NRA has reviewed the methods of colour measurement and the setting of standards, and concluded that the best approach appears to be based on the use of absorbance values [7]. Consequently the only sensible approach is to express standards as absorbance values over a range of wavelengths. Samples are filtered through a 0.45 μm filter, and absorbance in a 1 cm cell is measured between 400 and 700 nm. Standards are usually expressed at 50 nm intervals, although additional wavelengths may be used if sharp intermediate peaks are present.

The determination of the standards to be applied is also not straightforward. The usual approach for toxic substances, in which a safety factor is applied to known toxic concentrations, is not appropriate for colours. The problem is largely aesthetic and so there is a risk of subjectivity whatever approach is used. This can be overcome to some extent by using a panel to determine the standards, rather than relying on a single person.

Early attempts at setting standards were laboratory-based. Samples of effluent were diluted with river water until the residual colour was judged to be acceptable. Some variables, such as the depth of solution and lighting, needed to be tightly controlled in order to maintain consistency and this could be achieved in the laboratory.

The alternatives tried in the late 1980s involved either adding dyes to rivers or using the presence of coloured effluents in order to make a judgement about the acceptability relative to the measured absorbance of the water. These departed from the synthetic laboratory approach, and a panel could be used to make the judgements. There was, however, less control over such factors as depth of water, lighting and turbulence, so that the effects could be observed as different in adjacent river stretches. The research referred to earlier also involved field trials [7]. It was reported that there was a small boundary between what was acceptable and what was borderline. Water depth had an effect, as did background river colour and suspended solids. Comparing the results from different rivers in the East Midlands gave standards similar to those already being adopted by the NRA.

The methodology for the future is still under development, but is likely to be based on observations of rivers subject to colour pollution. This will involve taking observations and colour measurements at different sites and by different people over a period of time before arriving at a statement of what is acceptable.

There are natural variations in water due to the presence of dissolved humic and fulvic acids; waters derived from moorland peats have a yellow or brown coloration. Consequently standards based on aesthetic criteria are likely to depend on circumstances.

6 Current river colour standards in the Severn-Trent region

Despite determination at various times and by various people, the standards currently adopted for some of the rivers in the Severn-Trent region are very similar. Table 4 shows the standards used for determining consents for discharges in three catchments [13].

Table 4 Standards used for determining consents for discharges in three Severn-Trent catchments

River	Absorption at different wavelengths/nm						
	400	450	500	550	600	650	700
Churnet	0.025	0.015	0.012	0.010	0.008	0.005	0.003
Erewash	0.025	0.015	0.012	0.010	0.008	0.005	0.003
Bentley Brook	0.027	0.016	0.014	0.012	0.010	0.008	0.004

(handwritten annotations: LEEK → Churnet; PINXTON → Erewash; MATLOCK → Bentley Brook; DRABBLES)

7 Controlling discharges

The translation of river standards into limits to be applied in consents to discharge is straightforward. The usual method is a simple mass balance at each wavelength applied to the 95% exceedence river flow and the average effluent flow (Eqn 1):

$$A = \frac{(S \times T) - (F \times U)}{D} \tag{1}$$

where A = absorbance limit in discharge

S = river standard as absorbance

T = total flow (river plus discharge)

F = river flow upstream of discharge

U = absorbance of river water upstream of discharge

D = average flow of discharge.

The calculated absorbance would be applied as an absolute limit. This has already been done at a few sites.

8 Conclusions

The presence of coloured effluents in watercourses is no longer acceptable. The general quality of rivers has improved, and the wildlife, conservation and recreational value are increasing. Bright, unnatural colours are a cause of complaint and the public are increasingly critical of industry polluting the environment.

Colour standards are being developed for watercourses and being translated into limits for the control of discharges. The NRA intends to achieve a progressive reduction in colour pollution over the next few years by introducing new limits for the discharges not already subject to control of colour.

9 References

1. B D Waters, *Water Pollut. Control*, **78** (1979) 12.
2. B D Waters in *Surveys in industrial wastewater teatment*, Vol. 2, Ed. D Barnes, C F Forster and S E Hrudey (London: Pitman, 1984) 191.
3. G Morris and B D Waters, Proc. symposium on textile industry trade effluents, Institution of Water and Environmental Management, London (1993).
4. *River water quality – the next stage* (National Water Council: 1977).
5. *The quality of rivers, canals and estuaries in England and Wales – report of the 1990 survey*, Water Quality Series No. 4 (National Rivers Authority, 1991).
6. *The quality of rivers, canals and estuaries in England and Wales (1990–92)*, Water Quality Series No. 19 (National Rivers Authority, 1994).
7. P J Brown *et al.*, *Colour standards for watercourses* (National Rivers Authority, in press).
8. *River quality – the government's proposals: a consultation paper* (London: Department of the Environment, 1992).

9. The Surface Waters (River Ecosystem) (Classification) Regulations, Statutory Instrument No. 1057 (London: HMSO, 1994).

10. Regulations being prepared to implement 78/659/EEC.

11. Surface Water (Classification) Regulations, Statutory Instrument No. 1148 (London: HMSO, 1989) (implements 75/440/EEC).

12. Surface Waters (Dangerous Substances) (Classification) Regulations, Statutory Instruments Nos. 2286 (1989) and 337 (1992) (London: HMSO).

13. R McLellan, Paper to IWEM Jenkins Memorial Competition (Dec 1992).

CHAPTER THREE

The water company's view

John H Churchley

1 Introduction

Since the implementation of the Drainage of Trade Premises Act 1937 it has been common practice for most inland trade effluents to be discharged to foul sewer. Treatment of the trade effluent in admixture with domestic sewage then takes place at the sewage treatment works and the treated effluent is discharged to the watercourse. The textile dyeing and finishing industry has followed this general trend, and it is accepted that the large majority of dyers discharge to foul sewer [1].

In general this co-treatment of industrial waste water with domestic sewage has several advantages, especially where the major component of the chemical oxygen demand (COD) is biodegradable and the minor components are not detrimental:

(a) the domestic sewage provides pH buffering for the trade effluent

(b) inhibitory materials will be diluted by the domestic sewage

(c) domestic sewage provides nutrients such as nitrogen and phosphorus which may be insufficient in the trade effluent for biological treatment.

Sewage works effluent consent conditions include limits for biological oxygen demand (BOD), suspended solids and frequently ammonia. These may be achieved by conventional sewage treatment processes such as sedimentation, biological oxidation (biofilters or activated sludge plant) and tertiary polishing. Trade effluent dischargers are charged for waste water treatment on the basis of a modified Mogden formula [2].

For parameters that are poorly removed on conventional sewage treatment, however, the discharge to foul sewer followed by 'treatment' at the sewage works simply serves to dilute the material and provide dispersion into the aquatic environment. Where these materials are deemed to cause nuisance at the dilution afforded, then specific standards for the treated sewage effluent discharge are likely to be set by the National Rivers Authority (NRA). To meet these standards in its sewage effluent the water company can either:

(a) apply a specific nonconventional process on the sewage treatment works to remove the pollutant, or

(b) restrict the concentration of the pollutant in the trade effluent so that the sewage works effluent complies with its discharge consent.

Colour in effluent (from dye waste residues) is generally incompletely removed in conventional sewage treatment processes. This chapter describes ozone treatment for colour removal at Leek Sewage Treatment Works, as well as polyelectrolyte trials at the Leek, Wanlip (Leicester) and Pinxton works. Potential charging methods are also discussed.

2 Leek Sewage Treatment Works

2.1 Background

The town of Leek has been a centre for textiles production since the rise of the silk trade in the 18th century. By the mid-1970s, however, the Leek textile industry was largely confined to the dyeing and finishing of knitted goods.

At this time, the dyeing processes were giving rise to highly coloured waste water which was discharged to foul sewer and treated in admixture with domestic sewage at the sewage treatment works in Leekbrook. The colour was inadequately removed by conventional sewage treatment and gave rise to unacceptable coloration of the River Churnet, into which effluent from the Leek treatment works discharged. This gave rise to the setting of a colour discharge standard for the sewage works effluent, and after researching various options, led to the adoption of pH correction followed by alum dosing of the incoming mixed sewage [3].

For some years this situation was satisfactory, and adequate colour removal was achieved by chemical coagulation and conventional biofilter treatment. By the late 1980s, however, many of the Leek dyers had adopted the use of reactive dyes, and the nature of the dye waste received at the Leek works had changed accordingly. This had the effect of increasing the quantity of colour discharged and also of decreasing the treatability by chemical coagulation. These two effects caused increased levels of colour to be discharged to the River Churnet.

In 1989, the newly formed NRA required a more stringent standard to be met by the sewage works effluent. The new standard required a tightening of the BOD_5 (ATU)[1], suspended solids, ammonia-nitrogen and colour consent conditions. These new conditions were applicable from 1 April 1992. Old and new consent conditions are shown in Table 1. The new sanitary consent conditions required provision of increased capacity on the sewage works. The considerable tightening of the absorbance values gave most concern and prompted considerable research effort. The high proportion of dye waste (60%) in the total flow meant that colour removal on the sewage works was feasible.

Table 1 Consent condition for Leek STW effluent before and after 1 April 1992

Parameter	Before 1 April 1992	After 1 April 1992
BOD_5 (ATU)	35 mg l^{-1}	20 mg l^{-1}
Suspended solids	45 mg l^{-1}	30 mg l^{-1}
Ammonia-nitrogen	10 mg l^{-1}	5–10 mg l^{-1} [b]
Absorbance[a] at:		
400 nm	0.15	0.060
450 nm	0.12	0.040
500 nm	0.12	0.035
550 nm	0.12	0.025
600 nm	0.10	0.025
650 nm	0.08	0.015

a Absorbance measured in absorbance units in a 10 mm cell after pre-filtration through 0.45 µm membrane
b Lower figure in summer; higher figure in winter

2.2 Measures taken

A review of the international technical literature revealed that very little information had been published on full-scale colour removal to the standards required by the NRA. Chemical coagulation of reactive dye residues using inorganic salts was investigated using jar tests and found to be inadequate, and full-scale experiences at Leek supported this. Chlorine was shown to be effective on all but direct and disperse dyes and was six times cheaper than ozone [4]. Because of the risk of formation of persistent and highly toxic organochlorine compounds by chlorination of the industrial sewage effluent, this treatment was discounted for use at Leek.

---

[1] Biological oxygen demand, five days, with allylthiourea added to suppress nitrification

Biological treatment of dyes under aerobic conditions is slow, and enhancement of aerobic biological treatment was thought unlikely to achieve adequate colour removal [5]. Adsorption of reactive dyes on activated sludge was reported to be poor, but some removal of other dyes may be expected [6].

Anaerobic biological decolorisation of azo dyes was reported, and it was assumed that reactive dyes of this structure class would be adequately decolorised [7,8]. This treatment was rejected, however, because of the probability of odour problems from formation of reduced sulphur compounds, especially where sodium sulphate is used as dyebath additive. Chemical reduction was thought likely to suffer from the same disadvantage.

Adsorption of dye residues on activated carbon was found by Porter to be good [9], though the type of dye used in the study was not stated. Although costs were thought to be high, it was decided to investigate the use of activated carbon further. Similarly ozone was thought to be expensive but reports of its efficacy on a range of dye wastes looked sufficiently promising to warrant further investigation [4,10,11].

Pilot-scale trials were carried out: firstly, using packed granular activated carbon (GAC) columns and, secondly, using ozone generated from air. Severe fouling of the GAC columns occurred rapidly and even once-daily backwashing was inadequate to prevent plugging. In spite of the continuous fouling, it was apparent that the capacity of the carbon for colour removal was very limited and that treatment costs would be extremely high. Recent trials at another location in Severn Trent Water using a continuously washed GAC column have confirmed the very high costs for colour removal.

Pilot work with ozone treatment of the coloured sewage works effluent showed ozonation to be a powerful technique for colour removal, confirming the early work by Waters at Hinckley Sewage Treatment Works [12]. Reactive dye residues were effectively treated to the absorbance values required by the new NRA consent. Extensive pilot-scale experiments over a six-month period led to the conclusion that the 70% colour removal required at Leek was achieved by an ozone dose of 9.5 mg l^{-1} in a contact period of 20 minutes. These data formed the basis of the full-scale ozone plant design and are reported more fully by Gough [13].

As a result of the pilot study, ozonation was selected for full-scale colour removal at Leek. In order that the ozone dose should be minimised, it was decided to provide balancing in lagoons and effluent polishing in the form of sand filters upstream of ozone treatment. The

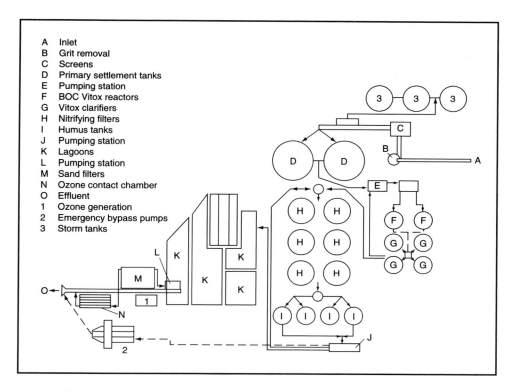

A   Inlet
B   Grit removal
C   Screens
D   Primary settlement tanks
E   Pumping station
F   BOC Vitox reactors
G   Vitox clarifiers
H   Nitrifying filters
I   Humus tanks
J   Pumping station
K   Lagoons
L   Pumping station
M   Sand filters
N   Ozone contact chamber
O   Effluent
1   Ozone generation
2   Emergency bypass pumps
3   Storm tanks

Figure 1 *Leek WRW schematic process flow chart*

tightening of the effluent sanitary consent conditions (Table 1) required the provision of extra biological capacity which was provided as a Vitox oxygen activated sludge plant. In total the works extensions amounted to a £10 m scheme of which the colour removal processes (lagoons, pumping station, sand filters, ozone treatment) cost approximately £5.1 m in 1990. Figure 1 shows a plan of the extended Leek site.

Ozone is generated from air in two generators, each producing up to 7.5 kg h^{-1} and delivering air containing ozone at a concentration of up to 22.9 g m^{-3} to a contact chamber comprising four lanes each of four compartments. In the contact chamber, ozonated air is discharged via ceramic diffusers into the sewage effluent. The concentration of ozone applied to the waste water is varied manually to achieve the required colour standard.

The works extensions were completed in October 1992 and the various plant items have in general performed well since that date despite some post-commissioning problems. Further

Table 2 Leek STW effluent quality/absorbance values before and after completion of extensions

| Wavelength /nm | Absorbance values in 10 mm cell | | | |
| | Before extensions[a] | | After extensions[b] | |
	Mean	Maximum	Mean	Maximum
400	0.055	0.199	0.037	0.074
450	0.048	0.175	0.022	0.054
500	0.044	0.174	0.016	0.046
550	0.023	0.100	0.011	0.030
600	0.016	0.068	0.0066	0.022
650	0.0089	0.034	0.0047	0.015

a 1 August 1991 to 31 August 1992; 49 samples, 11 consent failures
b 11 October 1992 to 10 June 1994; 83 samples, 5 consent failures

details of the plant and extensions and early operating experiences are given by Churchley *et al.* [14].

2.3 Performance

Colour removal performance for the extended Leek works is summarised in Table 2. For the 13 months period prior to commissioning the works' extensions, the mean absorbance for the final effluent ranged from 0.055 at 400 nm to 0.0089 at 650 nm. In this period there were eleven consent failures from 49 samples taken. After commissioning the extended plant, the mean values for absorbance fell to 0.037 at 400 nm ranging to 0.0047 at 650 nm. The rate of consent failures fell to five (measured against the new tighter consent) from 83 samples; since commissioning

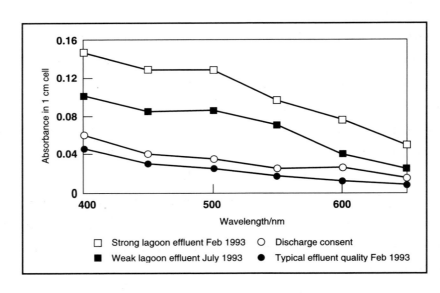

Figure 2 '*Strong and weak' lagoon effluent, colour discharge consent conditions and typical effluent quality at Leek STW*

there have been no reported complaints of colour downstream and the effluent has shown a marked visual reduction in colour.

Typical monthly average absorbance/wavelength plots are shown in Figure 2 for strong and weak lagoon effluent and for final effluent after ozone treatment. Reduction in absorbance from 0.15 to 0.048 at 400 nm is achieved by ozonation.

Whilst colour removal performance has been good, certain modifications will be made to improve future performance and to reduce ozone usage. These include modifications to the ozone diffusers, and automatic alteration of ozone dose in response to changing lagoon effluent colour.

3 Chemical coagulation

3.1 Introduction

The use of chemical coagulation for textile waste water treatment is well known, with the coagulation and separation process being applied either before or after biological treatment [15]. In the UK, chemical coagulation/flocculation has been used for the purification (including colour removal) of dye waste discharged to river [16–18]. Factors likely to be important in determining the success of these colour removal processes are adequate balancing facilities and the ability to prevent usage of certain troublesome dyes.

At the sewage treatment works, the dye waste is diluted with domestic sewage. The colour is thus present in lower concentration, and pH changes to promote flocculation are much less practical in the larger volumes of mixed waste water. Several dyers may discharge into the sewerage system and it may prove impossible to identify the source of any particular dyes giving treatment problems. The use of inorganic coagulants on dye wastes in admixture with domestic sewage is likely to be impractical owing to the high doses required, the large volumes of sludge produced and the risk of floc carry-over on to biofilters. The use of organic coagulants at much lower concentration and with much lower sludge production offers an attractive treatment option which is potentially easily retrofitted on a conventional sewage treatment works.

Chemical coagulation using organic coagulants has been investigated at Severn Trent Water over several years using jar tests. Claims of low sludge production and good colour removal have in general been substantiated, even on sewages containing predominantly reactive dye residues. Nevertheless, some products that successfully remove colour at one location appear ineffective at another site. This leads to the view that a particular product may be effective on one dye type or mix and that changes in the dye residue may render treatment ineffective. Even so, successful jar tests have led to the use of full-scale or part-scale trials at certain locations. Brief descriptions of unsuccessful and successful trials are given below.

3.2 The Leek experience

Although the ozone treatment described earlier was successful at removing colour to the NRA consent level, it was felt that the use of chemical coagulation using an organic polymer would have specific benefits.

(a) The ozone dose could be reduced if partial colour removal was achieved upstream. There would be a cost saving if the reduction in ozone costs outweighed the polymer costs.

(b) In the event of breakdown of both ozone generators the use of chemical coagulation would enable the works effluent to remain compliant with the colour consent conditions.

Jar tests identified an organic coagulant supplied by Nalco UK Ltd as being highly effective. This product was dosed into the sewage flow at various points in the treatment plant over several months. Initially, the dose was applied to the raw sewage before the primary sedimentation tanks. Secondly, the polymer was applied directly to the Vitox activated sludge plant reactors where mixing was excellent. Finally the feed to the humus tanks was dosed at a point of good mixing.

With all three dosing points, colour removal achieved was much less than jar tests predicted and at no time during dosing could the ozone dose be reduced.

3.3 The Pinxton experience

Pinxton Sewage Treatment Works receives dye waste from a single dyer; the dye waste flow amounts to approximately 40% of the total sewage flow. The dyer discharges predominantly direct and disperse dye residues, and the sewage effluent causes considerable coloration in the

River Erewash. This coloration has led to complaints by the public and by anglers, and as a result colour consent conditions have been set [19].

The works is atypical in that the dye waste is received in a separate sewer. When an extension to the works biological capacity was required in 1993, the cheaper capital option was biological pretreatment of the dye waste. This was achieved using a BOC Vitox activated sludge plant commissioned in July 1993.

Colour removal using organic polymers added to Vitox mixed liquor samples was investigated at jar-test scale. An organic polymer provided by Allied Colloids Ltd proved to be effective. Information on toxicity (including toxicity to nitrification) gave no cause for concern at the expected concentrations in the biological filter feed. A full-scale trial was arranged, with dosing of the organic polymer into the return activated sludge pipe at a point of reasonably high velocity just ahead of the Vitox reactors. Observed colour reduction across the Vitox plant was poor, however, and was far less than that observed in the jar tests. More importantly, however, there was a rapid rise in the ammonia concentration in the works final effluent from the second day of the trial. The rise in ammonia was accompanied by a drop in total oxidised nitrogen, indicating inhibition of the nitrification process. As a result the works final effluent failed the ammonia consent condition on two NRA samples. The dosing trial was abandoned and nitrification was slowly restored. Subsequent nitrification inhibition tests indicated that the polymer used was much more inhibitory to nitrification than had been indicated by the supplier [20]. To date the discrepancy between the suppliers and the water company's toxicity data has not been resolved.

3.4 The Wanlip experience

Wanlip Sewage Treatment Works serves the city of Leicester and treats a mean daily flow of 130 000 m³ per day. Approximately 11% of this flow is derived from some 45 dyehouses. Other significant industrial waste waters treated include those from potato crisp manufacture. The colour of the treated waste water often gives rise to coloration of the River Soar, and the NRA have given notice of absorbance standards to be met by January 1996.

Jar tests carried out on the activated sludge plant mixed liquor and using Magnafloc 368 supplied by Allied Colloids Ltd were successful in removing colour down to the concentrations required by the future NRA consent. Accordingly plans were made to carry out a part-scale trial using Magnafloc 368. The existing sewage effluent consent conditions required a high degree of nitrification, and so it was decided to carry out some on-site nitrification toxicity tests on the polymer. Two mini-biodisc units were set up and were fed on Wanlip settled sewage at a loading conducive to nitrification. When nitrification had been achieved reliably in both units, they were operated for two weeks with daily sampling of influent and effluents. Once it was established that there was no difference in performance between the biodisc units, a dose of Magnafloc 368 was applied to the settled sewage feed to the 'test' biodisc. The other biodisc unit was operated on plain settled sewage and constituted the 'control'. At 5.0 and 10 mg l^{-1} Magnafloc 368 there was no inhibition of nitrification in the 'test' unit compared with the 'control'.

A dosing trial was arranged with Magnafloc 368 solution (1% strength) dosed into two of the 16 lanes of the old activated sludge plant for a 19-week period. The two lanes chosen were lanes 15 and 16, which nitrify and have dedicated clarifiers and a separate sludge recycle. The chemical was dosed into pocket five of the ten activated sludge pockets to give an initial polymer dose of 2.0 mg l^{-1}, based on the settled sewage flow. Later this was increased to 4.0 mg l^{-1} and eventually to 5.0 mg l^{-1}.

At dose of 4.0 mg l^{-1} the colour removal on the dosed lanes was visually better than on the undosed lanes. Absorbance measurements confirmed this improvement, though the absorbance levels achieved did not match the proposed NRA consent conditions [21]. Table 3 shows the consent conditions that were met at 100% compliance compared with the proposed NRA consent and the absorbance values in the undosed control lanes effluent.

Table 3 Magnafloc 368 dosing at 4.0 mg l^{-1}

	Absorbance values in 10 mm cell at different wavelengths/nm						
	400	450	500	550	600	650	700
Undosed lane effluent (mean)	0.051	0.042	0.042	0.039	0.037	0.018	0.003
Lanes 15 and 16 effluent (mean)	0.030	0.022	0.018	0.015	0.015	0.007	0.002
Lanes 15 and 16 absorbance values met at 100% compliance	0.052	0.043	0.041	0.042	0.036	0.019	0.005
Proposed NRA consent (100%)	0.035	0.023	0.020	0.021	0.012	0.012	0.008

Table 3 shows the significant colour difference between the undosed lanes and lanes 15 and 16. The absorbance values met at 100% compliance for lanes 15 and 16 are shown to range from 0.052 at 400 nm to 0.005 at 700 nm; these are well outside the values for the proposed NRA consent.

Towards the end of the trial the dose level was increased to 5.0 mg l^{-1} for a short period. During this time wet weather conditions led to a large dilution of the incoming colour, so that little may be concluded from this phase of the trial.

To achieve the NRA consent conditions for absorbance on the required 100% basis, a dose level significantly above 4.0 mg l^{-1} will be required. The effects of such a dose on sludge settleability and sludge production are untested.

3.5 Conclusions

Chemical coagulation/flocculation as a process for colour removal has several attractions, especially when organic coagulants are used. Principal amongst these is the low sludge production and apparent ease of retrofitting to an existing conventional sewage treatment plant. Jar-test results indicate excellent colour removal for most dye types, including reactive dyes.

Jar-test performance has been difficult to achieve in either full- or part-scale trials. The reason for this is uncertain but may relate to inadequacies in mixing and flocculation conditions in the full-scale plant. The changing nature of the received dye residues and the varying flow rate may also be relevant factors.

This inefficiency in performance of the polymers at full scale may have serious repercussions. Most of the polymers used for colour removal are cationic and may be toxic to fish at low concentration: Wolf quotes LC_{50} values for freshwater fish species of between 0.09 and 0.40 mg l^{-1} [22]. Since the polymer residues in effluents cannot be measured directly at these concentrations, less efficient colour removal may lead to toxic concentrations of polymer undetected in the sewage effluent.

Many of the organic polymers offered for colour removal have an adverse effect on nitrification at likely dose levels [23]. Inadequate toxicity data provided by the polymer suppliers is likely to compound this problem. At most sewage treatment plants where nitrification is important, it will be necessary to carry out nitrification inhibition tests on any products to be used. Clearly the use of inhibitory products for colour removal at the dyehouse is likely to give rise to problems on the sewage treatment plant to which the dye waste drains.

The use of organic polymers offers a potentially attractive short-term amelioration of the colour problem. Serious problems of nitrification inhibition and fish toxicity can arise, however. It seems hardly rational to solve an aesthetic problem by dosing polymers at concentrations many times greater than the toxic threshold when no reliable method of measuring the residual concentration is available.

4 Charging for colour and colour removal

The possibility remains that sewage treatment works may provide or contribute to a practical and economic solution for colour removal. But with the wide range of conditions encountered, particularly in respect of dye types, seasonal and fashion production cycles and percentage of coloured effluent in admixture with sewage, it is unlikely that a single solution will be developed.

The adoption of colour removal at sewage works has to be dependent upon there being no detrimental effect on traditional sewage treatment processes or on the receiving watercourse, either by the effluent discharge or any storm water provisions.

Moreover, should colour removal be undertaken, it will be in addition to the usual sewage treatment processes on which trade effluent charges are based. Inherent in the provision of a colour removal scheme will be an additional charge to cover these additional pollutant-specific costs.

5 References

1. S J Hobbs, *J.S.D.C.*, **105** (1989) 355.
2. R C Bolton and L Klein, *Sewage treatment – basic principles and trends*, 2nd Edn (London: Butterworth, 1971).
3. N Harkness and J D Jedson, Severn Trent Water Authority R & D project report RT19 (1975).
4. W S Perkins, J F Judkins and W D Perry, *Text. Chem. Colorist*, **12** (1980) 182.
5. U Meyer, *Proc. FEMS Symp.*, (1981) 371.
6. H R Hitz, W Huber and R H Reed, *J.S.D.C.*, **94** (1978) 71.
7. *Dyes and the environment – reports on selected dyes and their effects*, Vol. 1 (American Dye Manufacturing Institute, Sept 1973).

8. D Brown and P Labourer, *Chemosphere*, **12** (1983) 397.

9. J J Porter, *Amer. Dyestuff Rep.*, **61** (Aug 1972) 24.

10. C Nebel and L Stuber, *Proc. 2nd Int. Symp. on Ozone Technology*, Montreal (1976) 336.

11. J M Green and C Sokol, *Amer. Dyestuff Rep.*, **74** (4) (Apr 1985) 50, 67.

12. B D Waters, *Water Pollut. Control*, **78** (1979) 12.

13. A J Gough, *Proc. Int. Ozone Assoc. Conf.*, Berlin (Apr 1993).

14. J H Churchley *et al.*, *Proc. ICE/IWEM Conference*, London (Mar 1994).

15. P Grau, *Water Sci. Technol.*, **24** (1) (1991) 97.

16. E J Newton, *J.S.D.C.*, **109** (1993) 138.

17. *Ind. Waste Management*, (Nov 1992) 22.

18. B W Iddles, *Proc. Textile Institute Conference*, Nottingham (Oct 1993).

19. R McLellan, Paper given to IWEM West Midlands branch meeting, Lea Marston (Dec 1992).

20. A Collins, Severn Trent Water Authority R & D report (June 1994).

21. J Jeavons, Severn Trent Water R & D report (Sept 1993).

22. C H Wolf, MSc thesis, University of San Francisco (1983).

23. A Collins, Severn Trent Water R & D report (Apr 1994).

CHAPTER FOUR

The trade association's view

John Harrison

1 The politics of colour

This chapter provides an insight into the prolonged and still-continuing political lobbying campaign in which the Knitting Industries' Federation (KIF) has been engaged following the declared intent of the National Rivers Authority (NRA) to introduce, for the first time, colour consents on dyehouse effluent to sewers.

In February 1990, a meeting in Nottingham of the KIF Dyeing and Finishing Sector was called at the request of Severn Trent Water (STW). Against the background of a claimed increase in the number of complaints from sectors of the public, STW explained that the then recently formed NRA, the government's new watchdog on water quality, intended to introduce consents on the discharge of colour in textile effluent received at water treatment works. Treatments at sewage works, it was reported, were ineffective at removing certain dyes and, in particular, (though not exclusively) reactive dyes. The colour was, therefore, passing through the treatment processes and into rivers. Should an NRA consent be applied, STW would become exposed to prosecution.

According to STW, the problem was materially and predictably exacerbated where there was a concentration of dyehouses; the volume of the receiving water was relatively low and therefore the level of dilution less, consequently increasing the propensity of the colour to be visible. The situation was thus more acute for those dyers who were discharging ultimately into the comparatively narrow, shallow and slow-flowing River Soar, which runs through the county of Leicestershire. The treatment works at Loughborough, Hinckley, Wigston and the large site at Wanlip, which serves the city of Leicester, were all identified as problem areas. A stretch of the Erewash River at Pinxton was also being targeted. Problems on the River Churnet, at Leek, Staffordshire, had been discussed late in 1989.

Astonishingly, STW was not prepared to offer any solution itself, but laid responsibility for the resolution of the problem firmly on the shoulders of the individual dischargers.

2 Closing the ranks

The KIF lost no time in forming an Effluent Working Party, which was rapidly renamed the Environmental Committee in order to reflect its quickly established reputation, status and widening responsibilities.

It was soon confirmed that there was no known viable end-of-pipe solution for the overwhelming majority of dyers who discharged direct into sewers. It quickly emerged, in the view of the industry, that the water company on technical, logistical and cost grounds was materially better positioned to deliver a solution, at least in the short term, pending the outcome of the recognised search for end-of-pipe solutions.

Next, it became clear that the proposed colour consents did not have their origin in any specific national or European Community legislation but in the general statutory powers of the NRA as laid down by the Water Resources Act. Paradoxically, the colour in textile effluent was nontoxic at the levels discharged. But each of the potential solutions examined (even if they were viable) had, in varying degree, an adverse environmental impact in increased energy consumption and/or the toxicity of by-products.

Worst of all was the threat to the livelihoods not only of the 4000 people employed directly in the affected dyehouses but also of those workers engaged in the manufacture of textile fabrics and garments in the East Midlands region. It was estimated that, without the essential capability to carry out the coloration of yarns, fabrics and garments, as many as 40 000 jobs were at risk, representing around one in four manufacturing jobs in the region and as many as one in two jobs in the inner city of Leicester. The threatened economic and social dislocation and misery was unthinkable at a time of mounting unemployment, together with the major contraction of the important coal mining industry in north Nottinghamshire, Derbyshire and Leicestershire, widely foreseen and hanging over the East Midlands' economic and social well-being like the sword of Damocles.

The KIF soon discovered that similar approaches were being made by North West Water to the Lancashire cotton dyers and finishers, represented by the Textile Finishers' Association (TFA). The TFA had commissioned March Consulting to carry out an environmental study, with part funding from the Department of Trade and Industry (DTI), which could include the colour problem. The principal and immediate concern in the North West was the presence of pentachlorophenol (PCP) in effluent, following processing in the UK of imported woven cotton fabrics and yarns from countries such as India and Pakistan. March Consulting gave a presentation to the KIF Dyeing and Finishing Sector Committee in May 1990, and a decision was taken to participate in the project as colour moved higher up the agenda. KIF members were asked to contribute towards the cost. Following a world-wide trawl to identify practical solutions, March Consulting submitted their final report to a meeting of KIF and TFA members, in Wakefield, on 21 November 1990. The outcome was that the KIF agreed to support the TFA in its lobbying over the PCP problem, but would continue to lead the campaign over the problem of colour which was of paramount concern to its members. Meanwhile, March Consulting's global search had been unable to identify viable end-or-pipe solutions for those compelled to discharge into sewers.

Central to the KIF's declared strategic response was the need to raise the political profile. It was decided initially:

– to channel the political representation through the high-powered Chemical Industries' Association (CIA) Dye Maker/User Group
– to maintain an ongoing dialogue with STW
– to seek meetings with the NRA and the Department of the Environment
– to consider involving major retailers and
– to request members to send copies of any colour consents to the KIF in order that a databank of knowledge could be established.

Following a series of informal meetings with STW during the late summer of 1991, the time seemed right to request a high-level meeting with the NRA. This took place in November 1991. The KIF demonstrated, firstly, that the lack of any technology for practical site colour removal inevitably demanded a lengthy timescale for improvements and, secondly, that setting unattainable consent levels would put many dyehouses out of business. It was also stressed to the NRA that the colour problem, which it was felt was not exclusively confined to reactive dyes, could only be removed at the tertiary stage. This would involve many small and medium-sized dyehouses, often with inner-city locations, installing an entire treatment plant.

They had neither the space nor the substantial capital required to fund such installations. Moreover, the KIF felt strongly that overwhelmingly the most cost-effective solution was for the water companies themselves to remove the colour.

By May 1993 the NRA had given dyers in Leek information on absorbance tests and the numbers involved. The KIF Dyeing Sector Committee decided that, in the absence of any meaningful progress elsewhere, the KIF itself should take up the task of political lobbying to avert the threat to the industry. No other group had the same level of knowledge of colour problems; the KIF enjoyed a long-standing high reputation for its political lobbying activities, and the timeclock was rapidly ticking away.

At a meeting with STW on 23 June 1992, the NRA's proposed colour standards were revealed for the first time, with the declaration that these would be imposed some time in 1993 and, taking the most optimistic view, by 31 December of that year. The proposals put forward were equivalent to a 'gin clear' discharge.

The KIF decided to request a tripartite meeting with both STW and the NRA. The initial objective, supported by STW following a high-level meeting in August at its Birmingham headquarters, was to put pressure jointly on the NRA to accept a minimum of five years' breathing space, during which an attempt could be made to overcome the world-wide technology void at viable cost. The possibility of a phased approach was advocated by STW, but this was met with reservation by the KIF, who believed that there was either a possible solution or none at all. STW, as had been feared, claimed that the problem of storm overflow by-passing any removal at treatment works prevented it on technical grounds from accepting the responsibility of attempting to remove the colour.

Meanwhile, it was decided to commission the then Trent Polytechnic to carry out tests on colour samples taken from a representative selection of dyehouses, so that the KIF could provide independent evidence of the improvements necessary. This revealed that a drastic sixty-fold dilution was necessary if the NRA's proposed consent standards were to be achieved.

3 The KIF's political lobbying strategy

The principal weapon in the KIF's political armoury was clearly the risk to the jobs of the 40 000 workers in the East Midlands alone, together with the prospect of the lost production simply being filled by imports from countries in the Far East and elsewhere, including other European counties, who at best paid scant regard to pollution control.

Within the KIF ranks it was believed that cost-benefit analysis, increasingly being advocated by the more pragmatic environmentalists, ought to result in a more reasonable stance being adopted by the NRA. Consultations with OFWAT, the water industry regulator, were deemed desirable. Background preparatory research revealed that by then the cost of the research being undertaken by many individual dyers, in association with a wide range of water treatment specialists, exceeded £300 000. A meeting with the Environment Minister, David Maclean, was requested. Meanwhile, it was considered important to continue the tripartite dialogue with STW and the NRA, and bilaterally, as necessary, with each of them.

At a tripartite meeting on 12 November 1992, it appeared that STW was again prepared to wash its hands of the problem: the STW simply restated that because of storm overflow the only solution open to them was to pass the NRA's consents on treatment works back up the pipe to individual dyers. The 'technology void at viable cost' argument was stressed repeatedly by the KIF, which believed that, in any event, STW was inherently in a materially better position in terms of existing primary and secondary treatment processes to deal with the colour problem at a tertiary stage. An unexpected advantage was gained at this meeting when the NRA representative expressed the view that storm overflow ought not to prevent STW treating the colour, adding that 'such overflow added beneficially to the level of dilutions' and, therefore, reducing the visibility of any colour present in the water.

At the meeting on 12 November, it appeared that the NRA was seeking help from both STW and the industry to ameliorate the immediate problem. The NRA recognised the nontoxic nature of the dyes, although it claimed that they might have an adverse impact on the flora and fauna in rivers because of the restriction they placed on the passage of light through the water, thus retarding photosynthesis. Meanwhile, the previously declared deadline of 31 December 1993 at the latest was emphatically restated.

At a meeting between KIF and OFWAT, it became clear that a request for an economic assessment of the severe adverse effects on the industry could be made to the Director-General of OFWAT, who could then appeal direct to the Secretary of State for the Environment.

Finally, a major new initiative emerged to set up a Tripartite Action Group, consisting of representatives from the KIF, STW and the NRA. Encouraged by recent statements from the

President of the Board of Trade, Michael Heseltine, about a renewed commitment by government towards industry, the KIF advocated the additional participation of the DTI which was subsequently accepted with observer status. The jointly agreed central objective was to produce an action plan which would attempt to enable the NRA to see positive ongoing improvements in the situation. STW announced that it intended to call a meeting of users in the Leicester area in December. The KIF Environment Committee met, in Loughborough, on 24 November to determine its approach at the Leicester meeting, together with the industry's input to the proposed action plan.

Meanwhile, the KIF, with the support of the All Party Knitting Industry Group at Westminster, had been instrumental in having several urgent parliamentary questions put down for written answer. The way was being paved for the anticipated KIF meeting with the Minister. Meanwhile, all members, along with the KIF itself, were encouraged to play their part whenever and wherever possible by explaining to the general public the nontoxic nature of colour in textile effluent and the global technology void in colour removal (other than for the handful of dyers with the space and capital to install their own miniature treatment works for direct discharge into an adjacent river). For those with no alternative to discharging into sewer, it now appeared that one apparently encouraging potential solution in the flocculation process could result in the residual polyelectrolytes adversely impacting on the nitrification process on domestic waste at treatment works. This would create increased emissions of ammonia, with an unacceptable potential lethal impact on fish and plant life in rivers.

It was also considered necessary to ascertain, with the co-operation of STW, the content of individual discharges bearing in mind the inevitable complexity of their composition and its seasonal variations, which arise from the extensive use of light pastel colours for spring/summer merchandise and the dominance of reds, blacks and navies for autumn/winter. The need for fair play between individual dischargers was also considered to be critical.

Television, radio and press interviews were organised, with the aim of articulating the threat in relation to the problem.

4 The political campaign takes a higher profile

The terms of reference for the Tripartite Action Group were finally agreed in January 1993:

–   to develop and implement a pragmatic action plan to ameliorate the colour problem at source and in river, with the aim of achieving acceptable river quality standards within an acceptable timescale

–   to collaborate in research to develop practical 'best available technology not entailing excessive cost' (BATNEEC) end-of-pipe solutions to fill the existing technology void

–   to maintain an effective public relations interface whilst the plan is being put into effect

–   to monitor the progress of the achievement of the action plan, and to update it

–   to act as a forum to resolve issues for all those discharging dye wastes to sewer (irrespective of their affiliation) and to encourage companies to exchange knowledge on best practice and to implement the findings.

By now the KIF was becoming increasingly anxious that an opportunity be provided to inform directly all dyers and finishers operating in the threatened areas, including the minority not in KIF membership. STW called a meeting of its Leicestershire-based dischargers, at which high-level officials from STW and the NRA gave audio-visual presentations to over 100 local dyers and finishers. These did little to allay the fears and frustrations of the audience. The KIF used the occasion to make clear its determination to protect the justifiable interests of the industry, within the policy of recognising that all dischargers of coloured effluent had a responsibility to contribute, individually and collectively, to the amelioration of the problem. A similar meeting was subsequently held in Mansfield for the Nottingham dischargers.

By the end of March 1993 it had become clear that the Tripartite Action Group had become deadlocked. The NRA was standing firm on its proposed consents and declared timescale. STW was claiming that, on the basis of its experience of ozone use at Leek, it was unable to meet the proposed standards, and therefore as a matter of policy was not prepared to treat colour at its treatment works.

Meanwhile, the threat to the very future of the industry remained. In a letter to the managing director of STW, Vic Cocker, dated 26 March, the KIF stated: 'We feel that we have given the tripartite process every opportunity but achieving the desired aim from our industry's point of view seems increasingly remote.' The letter continued by giving notice of KIF's intention to seek an urgent meeting with the Minister, with the key purpose of seeking support for persuading the NRA to postpone its declared operative date for the introduction of colour consents.

The support of the All Party Knitting Industry Group at Westminster, jointly chaired by Edward Garnier, MP for Harborough, and Alan Meal, MP for Mansfield, was now enlisted. KFAT, the industry's recognised trade union, was also invited to join the delegation because of the acute threat to jobs. As the prospect of appeals and associated legalities was looming ever larger, the KIF decided to consult a specialist law practice.

Following further dialogue in the Tripartite Action Group it was agreed to involve the dye manufacturers. The decision arose from the contribution made by the CIA representative at a meeting in May of the Leicester Action Sub-Group, at which the offer to provide a position paper had been readily accepted.

The principal elements to be found in the eventual CIA paper were:
- UK annual consumption of reactive dyes had peaked at the end of the 1980s at around 1200 tonnes, but had fallen back to around 1000 tonnes as a result of recession and the increased penetration into the country of low-cost imports of cotton leisurewear
- there was no early prospect of short-term solutions to the colour problem
- improved levels of dye fixation would only proceed incrementally.

A Joint Improvement Plan, involving the NRA, the water companies, the textile industry and the dye manufacturers was recommended, underpinned by a shared understanding of the constraints under which each party operated. Further initiatives on pollution prevention, as well as end-of-pipe treatment technologies, were also advocated.

In the meantime, a critical review of the available colour removal technology had been published [1]. The conclusions reached in this valuable review were that colour removal after biological treatment was feasible, but that the requirement to remove textile colour on site prior to discharge to sewer represented a 'technology void and/or a viability vacuum'.

5 The meeting with the minister

The KIF's attempts to meet with the Secretary of State for the Environment were frustrated because of his ultimate judicial responsibilities under the appeals legislation. As it became increasingly clear that neither the NRA nor STW seemed capable of being moved from their entrenched positions without intervention at the highest political level, a meeting was requested with the Parliamentary Under Secretary of State for Industry and Technology,

Patrick McLoughlin. The meeting was held in London, on 23 July 1993, with the KIF director leading the industry delegation which included Edward Garnier MP, a joint chairman of the industry's parliamentary lobby, and the General Secretary of KFAT, Helen McGrath.

A detailed KIF briefing note had been prepared in advance covering:

- the economic and social importance of the industry
- the serious threat and dilemma created by the NRA's proposed consents
- the world-wide lack of effective technology at viable cost
- the widespread efforts to find viable end-of-pipe solutions, involving investigations of several potential technologies, carried out by many dyers, at a collective estimated cost at that time in excess of £300 000.

The Minister shared the KIF's deep concerns but stressed that neither the DTI nor the Department of the Environment had any legal powers over the NRA. He suggested that the parties to the Tripartite Action Group should meet for high-level talks with senior officials in the DTI Environment Unit in an effort to find a workable solution through mediation.

The resulting meeting with DTI officials, the KIF and STW was held on 10 September 1993, in London. The TFA was also invited, together with the CIA's Dye Users' Group. The KIF restated its case in detail, including best estimates of the costs of carrying out treatment on sites, contrasting these with the much lower costs that STW was likely to incur if it undertook the responsibility.

STW reiterated that chemical additions (flocculation) carried potential serious risks to the sewage process. Therefore it would in the short term need relaxation of the proposed NRA standards, along with flexibility on the need to comply 100% of the time, without which STW would be exposed to prosecution. Moreover, STW believed that individual dischargers would need to contribute to the task by ensuring that effluent arriving at treatment works was better balanced.

The TFA reported that its DEMOS project, partly funded by the DTI, had been extended to embrace the colour problem. The findings (see Chapter 5, page 59) had confirmed that maximum balancing and maximum dilution facilitated colour removal, but that no end-of-pipe solution had been found at viable cost. The CIA repeated that the current research would take decades to achieve a level of fixation that did not involve visible colour in effluent and, in any event, progress would be incremental.

In conclusion, the KIF confirmed its willingness to meet again with the DTI and STW, to produce both short- and long-term action plans for submission to the NRA. It was agreed to hold a further meeting as a matter of urgency. This took place one week later on 17 September 1993, at STW's Birmingham headquarters, with the following terms of reference (proposed by the KIF to the DTI and agreed in advance by STW with minor modifications): 'The KIF and STW agree to produce an action plan to ameliorate, in the short term, the removal of colour from textile effluent and also a strategy for the long term. Solutions will be offered to the NRA by not later than 24 September 1993.'

Following further intense dialogue, during which the DTI's even-handedness between the parties never faltered, the KIF's persistence reaped the long-awaited but elusive dividend. STW finally agreed to undertake, on a trial basis and given reasonable legal protection, the responsibility of attempting to treat coloured effluent at selected treatment works; they stipulated that dyers should be prepared to collaborate, share best practice among themselves and also co-operate in an agreed long-term research programme. The catalytic role of the DTI had been significant; nevertheless, lingering doubts inevitably remained in the KIF.

Intense work on the production of the required action plan began at once. In the event the declared target date for completion proved too ambitious, but the plan was finally submitted to the NRA by the end of October 1993. Its principal aims were:

- pursuit of practical colour removal at selected treatment works
- examination of dyehouse discharge management procedures
- formulation of best practice for dyehouse operators
- establishment of further research and development programmes
- setting up of monitoring and review mechanisms.

The objective was to achieve an immediate reduction in the level of colour discharged into rivers following processing at treatment works.

The NRA welcomed STW's departure from its previous policy and gave its conditional support in principle, but with the critical reservation of its desire to see 'a peg in the ground' approach, i.e. no further deterioration from current levels. The KIF expressed sympathy with this concept, but nevertheless felt it necessary to articulate the inherent impracticabilities in terms of the base line and the seasonal and fashion variations inherent in a dynamic industry. The NRA remained adamant that 100% compliance must be achieved, and STW thereupon reverted to the policy of passing the problem back along the pipe to the dischargers. In a nutshell, if such attitudes remained rigid the industry was again back to square one. For some, faith in the political lobbying process was beginning to weaken. The negotiating process and frustrations had by then dragged on for three years and nine months, with the threat still very much present.

At this point the KIF and STW agreed to combine together with 'determined resolution' to attack the NRA's response. Meanwhile, the question was being addressed as to how best to spread the trial development costs incurred at the designated sites over the entire catchment area, within a framework that would stand up in law. The KIF's preferred approach was to defer charging dyers with the development costs at one treatment works until others came on stream (which should not be long delayed), when the costs could be shared more equitably. At the same time the KIF began promptly to discharge its accepted responsibilities within the action plan in respect of promoting enhanced dyehouse discharge management and establishing best practice for dyehouse operators. As soon as the application for the DTI's promised grant of £10 000 towards the estimated total cost of £30 000 had been approved, a specialist consultant was briefed and appointed. A full copy of the resultant KIF *Water and effluent management manual* was supplied to each member dyehouse following three audio-visual presentations to the industry, covering each of the affected catchment areas, during April and May 1994. Copies were offered to nonmembers at £275 each. The manual features:

— measurement systems and means of controlling the flow of water and colour in effluent
— information from seven dye manufacturers, over two liquor ratios, on how much colour will remain in discharged effluent
— a set of individual case studies covering all the major product sectors, i.e. garments/socks, knitted fabric with jet dyeing and mixed jet/winch, package-dyed yarn and ladies' hosiery.

Comparison of a company's own processes with the case studies enables it to determine whether it is adapting the optimum process and best housekeeping procedures in order to reduce the amount of colour in its effluent.

By the end of March 1994, it had become increasingly clear that the NRA was unlikely ever to give its formal approval to the action plan but, if all parties proceeded with their respective commitments, then the introduction of colour consents on the affected treatment works would be postponed until 1 January 1996. Meanwhile, however, the NRA insisted that some form of descriptive consents could be introduced. This was another material breakthrough, although its effect was only to buy time. The issue of the costs to the industry of the colour removal trials still remained unresolved.

The NRA by this stage had also demonstrated some flexibility on its 'peg in the ground' policy by acknowledging that this should not affect the normal dynamics of a fashion industry. Colour consents would, however, be immediately imposed on any new dischargers.

6 The task ahead

At the time of writing (the late summer of 1994) the agreed trials at STW treatment works are all still continuing, though not on a bulk basis. Liaison meetings between STW and the dischargers into each works are being pursued, and individual dischargers are utilising the KIF manual. The search for viable and effective end-of-pipe solutions continues.

With both water and effluent charges rising rapidly, the removal of the colour in dyehouses, and thereafter a capability to recycle the water, are clearly the ultimate goals. The PR profile has been maintained through participation in the BBC Radio 4 series *Talking Green Politics*, which dedicated a full 30 minute programme to the issue of colour in textile effluent on 13 August 1994.

Meanwhile, the collective support and participation of many members, together with the authoritative nature of technical input from industry experts, have resulted in the KIF's prolonged campaign gaining certain critical points and more time. The debate, however, goes on. The hope must be that commonsense will continue to prevail. No one can doubt the industry's ongoing resolution and commitment in terms of the time, financial resources and efforts which have been, and will continue to be, devoted to finding viable solutions.

If the KIF's fears are realised, however, and the STW trials fail to meet the set standard and viable end-of-pipe solutions remain elusive, the KIF will attempt to negotiate a more reasonable standard with the NRA in the short term. If all else fails, the KIF is prepared to co-ordinate, with maximum impact, the most appropriate of the possible legislative appeal procedures in an effort to preserve the industry's long-standing economic and social contribution to the well-being of the Severn Trent region and the nation at large.

7 Reference

1. P Cooper, *J.S.D.C.*, **109** (1993) 97.

Part Two   The search for solutions

Chapter 5

Industry evaluation of colour reduction and removal – the DEMOS project,
Barry G Hazel

The paper deals with the background to the Textile Finishers' Association DEMOS project and its aims and objectives before considering the current legal requirements for colour reduction and removal. It then looks at six technologies which were considered in detail, reporting on their efficiencies, capital investments, requirements and operating costs, and draws conclusions from the results of the project.

Chapter 6

Technical solutions to the colour problem: a critical review, *Timothy G Southern*

The technologies available to 'remove' colour are critically reviewed. The technologies for processing coloured effluents are subdivided into two main groups: those that simply concentrate the colour in either liquid or solid form and those that actually destroy the colour-containing components of the effluent. Within each main group, the technologies can be further subdivided into simple categories (such as membrane technologies, ion exchange, adsorbents), and the pros and cons of each technology examined. An assessment is made of the potential environmental impact, as yet unknown, of using partial oxidation techniques to eliminate the molecular structures that create the colour.

Chapter 7

The treatment of dyehouse effluent at Stevensons Fashion Dyers, *John Scotney*

The progressive development of an effluent treatment plant for a large commission dyer is described. A state-of-the-art biological treatment plant, installed in 1968, has been augmented during the last eight years to cope with extra production demands and the need for the removal of colour.

Chapter 8

The treatment of dyehouse effluent at John Heathcoat, *Eric J Newton*

This article describes the installation of an effluent treatment plant for a textile dyeworks to satisfy the regulatory requirements for both colour and biological oxygen demand for an effluent to be discharged direct to a watercourse.

CHAPTER FIVE

Industry evaluation of colour reduction and removal – the DEMOS project

Barry G Hazel

1 Introduction

At the end of 1989 the textile finishers in Lancashire were advised by Her Majesty's Inspectorate of Pollution (HMIP) of a major problem with textile finishers' effluent, in that it contained pentachlorophenol (PCP). This fungicide, a Schedule 5 (Red List) substance, was being added to fabric during the weaving stage, particularly in Asian countries, to prevent deterioration of the sizing bath and to prevent mildew in transit. When the fabric was scoured by the finisher the PCP was washed off into the effluent and was not being removed by the sewage works. It was appearing in natural waters, where it was prohibited under EC legislation.

Being unable to find a satisfactory technique for removing PCP from the effluent, the members of the Textile Finishers' Association, together with their colleagues from the Knitting Industries Federation, took the decision to employ the March Consulting Group to look for the best available technology for removing this product, as it was assumed that other countries had a suitable method of treatment. March reported back in 1990 that they had been unable to find any method of treatment on a commercial scale for the removal of PCP from textile finishers' effluent. In most countries, no attempts were even being made to detect its presence.

By the time of the first March report, the bleachers, dyers and printers of imported woven cotton fabrics were facing serious threats of legal action by the National Rivers Authority (NRA); indeed, two or three companies had already been prosecuted for the discharge of PCP in their effluent. It was therefore obvious that companies had to join forces to carry out their own research and develop their own systems. In collaboration with the March Consulting Group, a research programme was drawn up to cover waste minimisation and effluent treatment. At the time when the programme was proposed, the main effluent treatments being researched were the removal of chemical oxygen demand (COD), suspended solids and Schedule 5 compounds, particularly PCP, with colour as a minor item.

The research project was put forward to the Department of Trade and Industry for funding under the DTI's Environmental Management Options Scheme and was approved as a project under that scheme in December 1991. The lead collaborator in the consortium making up the project was the Textile Finishers' Association, with four collaborating sites selected to represent various types of printing, dyeing and finishing and, in addition, 36 subscribing companies. Where appropriate, small parts of the programme were carried out at some of these companies.

The waste minimisation research was extremely successful, with very large potential savings being demonstrated. The problems arising from the presence of PCP were largely removed by elimination at source. During the first part of the programme, however, the problem of colour became more and more apparent and much of the research effort was transferred to this area.

2 The textile colour-using industry

At the commencement of the research a study was made of the dyeing and printing industry. It was confirmed that most of the companies involved were small and medium-sized enterprises which, including those companies which were part of a group, usually operated on an individual site basis, and often as a profit centre. It was also found that over 90% of the dyers and printers discharged their effluent to a waste water treatment works (WWTW) for further purification. It is necessary to take these facts into account when considering treatment options.

3 Analytical methods for colour

Colour is measured as absorbency values (optical density) through a 10 mm cell at several wavelengths across the visible spectrum. Samples are filtered through a 0.45 mm filter before absorbency measurements are made. The results of colour analysis can be affected by using filter papers from different sources, and even different batches from the same source. Achieving consistency and repeatability has proved difficult.

The experimental work showed that there were unlikely to be narrow peaks across the wavelengths, and therefore measurement at a minimum number of wavelengths was recommended. There was also some concern about the effect of turbidity on the measurements.

4 Legal position

At the time of the research project the question of colour consent levels was principally at the discussion stage, although examples were found where the NRA had issued colour consent limits to a WWTW and also to dyers discharging direct to rivers. In the North West area the water authority had made certain dyers subject to colour consent limits on their discharge to sewer (Table 1).

The WWTWs measure the colour value of the total effluent received, regardless of the dye class, because it is not possible to distinguish, in an effluent, one type of dye from another. Most dye classes are primarily absorbed in the biomass in the sewage treatment works and mainly reactive dyes and certain acid dyes pass through the sewage works.

Table 1 Colour consent limits in North West England

| Source of consent | Colour standards for absorbance at wavelength/nm | | | | | | |
	400	450	500	550	600	650	700
To achieve RQO	0.025	0.015	0.012	0.010	0.008	0.005	0.003
NRA to WWTW	0.031	0.019	0.013	0.012	0.012	0.011	0.003
NRA to dyehouse	0.071	0.043	0.033	0.026	0.023	0.019	0.013
WWTW to dyehouse	0.4	0.3	0.06	0.10		0.10	0.10

The research for colour reduction was based on the three established principles, which are discussed in the following sections:

- elimination
- substitution
- treatments.

5 Elimination

Colour is the raw material for the dyer and printer, and cannot be totally eliminated if coloured textiles are to be produced. Based on the findings of the research, it may be possible for companies to make a decision to eliminate the use of certain classes of dye from their production.

The management of colour was also investigated and it was found that the use of automated weighing and automatic colour mixing, particularly in the printing companies, greatly reduced excess colour paste production and also resulted in less reprocessing. These techniques were less important for batch dyeing, and had less effect than in the continuous dyeing and printing area.

6 Substitution

There is a general perception that natural products are more environmentally friendly than synthetic products. This is certainly a misconception as far as dyes are concerned. Natural dyes have relatively poor fastness to both washing and light and are extremely difficult to fix to textiles. The process for this fixation produces a very polluting effluent and therefore these dyes were eliminated from the study.

It was found during the research that the shades and requirements for the dyed fabrics were often decided by designers who had little or no knowledge of the effect of their decisions in environmental terms, and it was recommended that environment studies should be included in designers' training courses.

7 Treatments

Studies were carried out of some thirty different treatments under the following headings, details of which are given in Table 2:

— coagulation, flocculation and precipitation

— oxidation

— adsorption

— electrolysis

— biological treatments

— membranes.

7.1 Coagulation, flocculation and precipitation

The inorganic coagulants – lime, magnesium and iron salts – have been used for coagulation of dye waste over many years. With changes in dyes, and with the dye consents proposed by the NRA, these no longer give completely satisfactory treatments.

Organic polymers have been developed for colour removal treatments and, in general,

Table 2 Summary of colour removal techniques

Technique	Key performance criteria	Capital cost /£ × 10³	Treatment cost[a] /£ m⁻³	Comments
Coagulation/ flocculation	Fails consent at 400 nm	100–300	0.14–0.20	Should be considered as first stage
Oxidation chlorine dioxide	Failed	15–60 (FOC lease possible)	2.80	
catalysed peroxide	Consent achieved	100–200	2.30–100	Should be considered for treatment of concentrated wastes
ozone	Consent achieved	350–1000	0.15–0.30	Treatment only after coagulation/ biological
Adsorption inorganic	Consent achieved	100–300	0.25–0.40	With after-treatment for recycling
bio	?	300–500	0.30–0.40	Recycling opportunities require checking
Electrolysis	Consent achieved	250–500	0.30–0.40	
Biological	Not reliable	200–2000	Low	Will require tertiary treatment for reliability
Membranes		Under development	0.5–20	Should be considered for final treatment to give possibility for recycling

a These costs assume a balanced effluent; many do not include sludge disposal they offer advantages over inorganics: sludge production is much less and colour removal was significantly improved. It was found, however, that colour consent conditions were not reliably met.

Cationic polymers, despite low mammalian toxicity, are toxic to freshwater fish at low concentrations and care must be taken that the flocculation conditions ensure low residual polymer. Consideration also has to be given to the sensitive sewage treatment processes, such as nitrification. Certain polymers may inhibit this process, giving rise to increased levels of ammonia in the river.

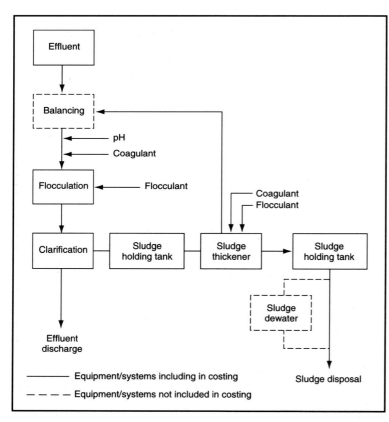

Figure 1 *Coagulation and flocculation flow diagram*

The trials carried out showed that consent conditions at 400 nm could not be met by coagulation and flocculation techniques, although others could be met. The pale yellow colour at this wavelength was not obvious to the eye. The conclusions concerning this technique were that it would work effectively, although not completely, to consent conditions on standard dyehouse mixed effluent.

Experimental work with mixed inorganic and organic coagulants, followed by tangential flow separation, showed promise; all consent conditions were not met, however. There was a requirement of space for the treatment plant (Figure 1) and the capital costs were high: from £100 000 to £3 million, depending on the size of plant required. The running costs for chemicals and utilities were of the order of £0.14 to £0.20 per m³. In considering this plant the cost of sludge concentration must be taken into account, as sludge disposal costs are growing rapidly.

7.2 Oxidation techniques

A variety of oxidising agents can be used to decolorise dye waste, and are described in turn below.

Chlorine

Chlorine, in the form of sodium hypochlorite, decolorises many dyebaths efficiently, though not all. It is a low-cost technique, but there is concern about any excess chlorine forming adsorbable organohalides (AOXs) with organic material in the effluent and the watercourses. In some countries, legislation on AOX levels is already in place.

Chlorine dioxide

Chlorine dioxide (ClO_2) is less reactive than chlorine and has been claimed to give rise to fewer side-reactions (Figure 2). The experimental study showed, however, that it did not decolorise dye waste efficiently to consent conditions, as it has no effect on some dye classes (such as vat dyes). Nevertheless, chlorine dioxide is highly effective against reactive, direct, disperse and anionic pre-metallised dyes. It could be used as a polishing treatment.

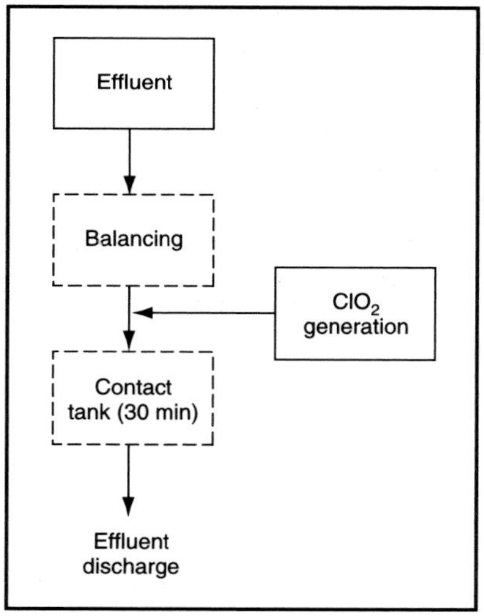

Figure 2 ClO_2 oxidation flow diagram (see Figure 1 for key)

Hydrogen peroxide

Hydrogen peroxide, alone, is insufficiently powerful to decolorise dye waste at a normal temperature and pH. In acid solution, however, with iron(II) (Fenton's reagent) as a catalyst, the peroxide forms the vigorous hydroxyl radicals and may be used to decolorise dye wastes.

Fenton's oxidation (Figure 3) is capable of treating both soluble dyes, such as reactive

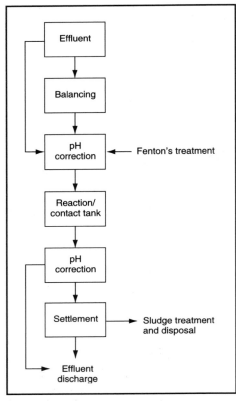

Figure 3 Fenton's oxidation flow diagram

dyes, and insoluble dyes, such as vat and disperse dyes, and did achieve colour consents for both the concentrated and the dilute wastes investigated. The vigorous oxidation also reduces the COD of the effluent. Neutralisation of the effluents after treatment causes precipitation of the iron oxide and hydroxide, which removes any remaining insoluble dye from the effluent by absorption and/or flocculation.

Fenton's oxidation was the most successful for treating concentrated colour wastes, such as pad dye wastes, but at an estimated cost of £80 to £176 per m^3. Treatment of these relatively small volumes gives significant reduction in the end-of-pipe colour and has potential as a first-stage treatment.

The cost of using Fenton's reagent for end-of-pipe treatment was estimated at £2.3 per m^3, but a COD reduction of 56% was achieved; however, this must be balanced against the costs for sludge disposal.

Ozone

Ozone is a very powerful oxidising agent which will decolorise dye wastes (Figure 4). A high COD significantly reduces the effectiveness of the colour removal, but is not itself reduced. Ozone must be generated on site and its toxicity to operating personnel must be taken into consideration. The capital costs are likely to be very high, in the region of £500 000 to £1 million, with running costs of the order of £0.15 to £0.30 per m^3.

Trials were also carried out with ultraviolet radiation/ozone and ozone/peroxide, but no appreciable improvements or cost benefits were determined over ozone alone.

The perceived advantages of oxidation techniques lay in the avoidance of sludge generation and the consequent removal and disposal costs. All techniques, however, either require a sludge-producing pretreatment (biocidal or coagulation, for example) or, in the case of Fenton's oxidation in which no pretreatment is required, produce a sludge as part of the process.

7.3 Adsorption

Activated carbon

Activated carbon is the original absorbent used to absorb dye molecules and remove colour. The activated carbon can be granulated (GAC) or powdered (PAC).

GAC is normally used in columns, through which the effluent is passed. GAC loses its capacity to absorb as the surface area for absorption becomes saturated with absorbed molecules. Regeneration is possible using heat, steam or chemical treatments.

Other chemicals in the waste water stream would also be absorbed in the column and thus blocking would occur very rapidly. This process could be considered as a polishing treatment, but is expensive in terms of both capital and running costs.

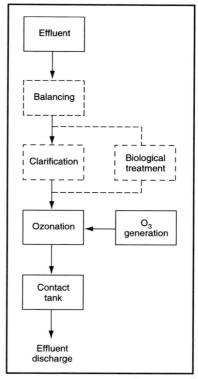

Figure 4 *Ozonation flow diagram (see Figure 1 for key)*

PAC is usually dosed as a slurry into the effluent stream for subsequent removal and disposal. Most dyes can be absorbed; contact times vary between a few minutes and an hour. A range of activated carbons is available; some are especially suited to a particular dye type.

Clays

Dyes can be absorbed on to several low-cost natural materials and special materials based on inorganic particulate synthetic clays have been developed for the adsorption of dye waste. The process required pH adjustment of the incoming effluent to 6.5–7.0, followed by treatment with the clay and then a further pH adjustment to 10 (Figure 5). Adsorbent clays were shown to reduce the colour significantly in the effluent, although not consistently to consent conditions.

Capital costs for equipment were between £100 000 and £250 000, with running costs for chemicals and utilities varying between £0.25 and £0.40 per m^3 when the reduction of COD and suspended solids had been taken into account. Regeneration and sludge disposal will again significantly increase these costs.

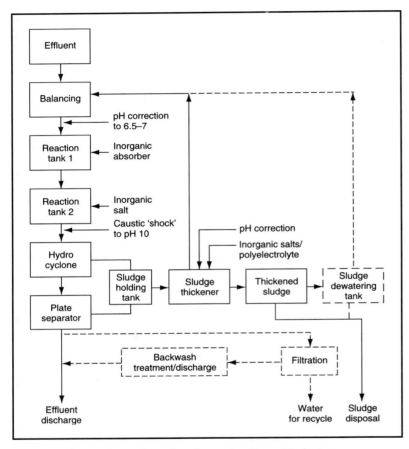

Figure 5 *Adsorption/ion exchange flow diagram (see Figure 1 for key)*

Bioadsorbents

Certain bioadsorbent products will adsorb dyes reversibly. The adsorbent is regenerated by an alkaline wash, which then contains the colour in a concentrated form, from which it must be coagulated and settled into a sludge for disposal. The effluent must be brought down to pH 5.6–6.0 before the adsorption and then brought back to 10 afterwards. At least one manufacturer recommends reverse osmosis at the end of the adsorption phase.

The project was unable to prove the process in full-scale trials or to prove that the consent levels could be achieved. Some concerns were expressed, however, particularly relating to salt concentration, which may affect the adsorbent performance, and to whether the polyelectrolytes would affect the reverse osmosis membrane performance.

Capital costs were estimated to be in the region of £3 million to £5 million, with running costs of approximately £0.50 per m^3, although this could be reduced if the opportunities for recycling are proven. The costs of sludge disposal must again be considered.

7.4 Electrolysis

The system depends on the electrolytic production of a flocculant from the aluminium of an electrode (Figure 6). This production of an aluminium flocculant in situ appeared to give better flocculation then the addition of inorganic chemicals.

Experiments were also carried out using electrolysis followed by the addition of an organic polymer. The electrolysis reduced the amount of organic polymer required, but showed no advantages in cost terms over organic polymer alone, although there may be slight improvements in flocculation.

The capital costs ranged from £250 000 to £500 000, with a running cost between £0.30 and £0.40 per m^3.

7.5 Biological treatments

Aerobic

During the standard biological effluent treatment colour is not destroyed, but considerable amounts may be adsorbed on to the biomass. This is dependent on the loading, and some breakthrough may be observed. As mentioned previously, this adsorption does not occur with reactive dyes or with some acid dyes; with these dyes, therefore, further treatment will be required.

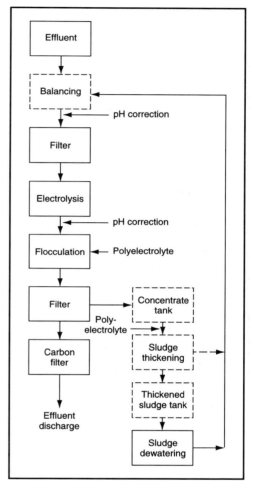

Figure 6 *Electrolysis/flocculation flow diagram (see Figure 1 for key)*

Anaerobic

Preliminary work on anaerobic degradation of effluent has indicated that it is not efficient for colour removal. Moreover, there is some concern that the presence of sulphates in the dye waste may give rise to hydrogen sulphide under anaerobic conditions.

7.6 Membrane techniques

The research investigated various types of membrane techniques that are curently commercially available. These are:

— microfiltration

— ultrafiltration

— nanofiltration

— reverse osmosis.

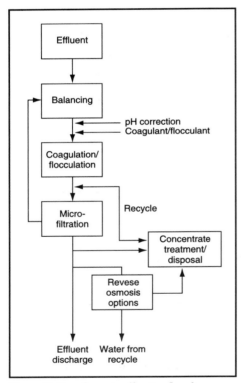

Figure 7 *Crossflow microfiltration flow diagram (see Figure 1 for key)*

Microfiltration

Microfiltration (Figure 7) separates materials in suspension from those in solution. It was considered in the project as a method of treatment after coagulation/flocculation. It has certain advantages, in that it copes with the effluents containing detergents and dispersing agents and it is relatively insensitive to temperature variations.

Microfiltration on its own is not a satisfactory end-treatment, but may be used for the partial removal of colour and organics before sewage discharge, or before reverse osmosis or activated carbon treatment.

Several types of microfilter are commercially available but have not been tested in the UK. Research by the Water Research Commission in South Africa has reported satisfactory results using tubular polyester units with a design pressure drop of 400 kPa.

Ultrafiltration

Ultrafiltration, a development of microfiltration, separates suspended matter and larger molecules in solution from the small molecules.

Nanofiltration

Nanofiltration retains most pollutants inside the membrane but allows water and some small-molecule salts and organics to pass through.

Reverse osmosis

Reverse osmosis allows water to pass through the system, but retains almost everything else from the waste stream. Conventional reverse osmosis (also known as hyperfiltration membrane technology) for treatment of industrial effluents has limitations based on physical conditions, such as pH and temperature, and chemical properties such as membrane–chemical interactions and fouling. The newer thin-film composites (TFC) are much improved in relation to these parameters. Most reverse osmosis membranes are manufactured for sea water desalination. Some are not stable to chlorine, and others are affected by cationic surfactants.

Two systems are available:
— use of TFC membranes in plate and frame or tubular form
— for hot dyehouse effluent (over 45 °C) only dynamic membrane technology, using stainless steel porous tubes, is effective.

Membrane technology is developing rapidly and companies are now taking more interest in effluent treatment.

8 Conclusions

No simple technique will remove colour from a complex dye waste mixture to meet consent conditions. A study of Fenton's oxidation showed that this technique could decolorise the dye waste investigated to consent conditions, but costs were high.

The process chosen must be matched to the dye waste of the particular factory at which it is installed. It must be recognised that the smaller the discharge, the higher the unit cost; most companies in the industry are small or medium-sized enterprises.

In many cases, a biological treatment or coagulation/flocculation treatment will be required before the tertiary treatment to remove colour residues. As most small finishing companies discharge their effluent to a sewage treatment works, the question must be addressed as to whether treatment for the removal of colour residues should be carried out at the treatment works. A company considering carrying out treatment would have to take into account the cost and space requirements for the primary and secondary stages, as well as for the tertiary colour removal stage. Most techniques will only perform on a balanced effluent and the cost of balancing tanks must be included.

From biological or coagulation/flocculation techniques large quantities of sludge will be developed, which will usually contain any toxic compounds that were in the original effluent. New legislation on sludge is likely to increase disposal costs substantially, and these costs must be considered when choosing a system.

Before investigating possible suitable treatments, companies must study in detail their total water usage and reduce this as far as possible. The DEMOS project showed significant opportunities for this reduction.

For some companies the combined cost of water purchase and effluent treatment already exceeds £250 000 per annum and these costs are increasing. A combination of the effluent treatment techniques described above gives the opportunity for recycling over 70% of the water. The advantages of this action will vary from site to site, but the possibility must be taken into account when considering medium- and long-term strategy.

CHAPTER SIX

Technical solutions to the colour problem: a critical review

Timothy G Southern

1 Introduction

The problems associated with the discharge of colour from dyehouses has concerned both industrial and academic scientists for at least four decades, not only in the United Kingdom but in most industrialised countries. The recent proposal by the UK National Rivers Authority to impose discharge consents for colour on discharges from water authorities has highlighted the need to identify economic treatment technologies to overcome the problem. In August 1993 the author presented a seminar to the members of the Knitting Industries' Federation that summarised the technologies used to resolve the colour problem [1]. Since that date, more open literature has become available and the author has been made aware of additional work being undertaken in various institutions.

While several articles dealing with this topic have appeared in the literature, most have concentrated on well-established effluent treatment technologies [2–5], or have presented the case from the administrator's point of view [6–9]. This article will, while referring to these established technologies, also examine those that are newer or less well known.

A reasonable idea of the effluent composition and required treated water quality is required if the problems associated with the aqueous discharges from textile dyers and finishers are to be effectively addressed. A recent review of the effluent discharges into a particular sewage treatment works has revealed that an 'average' effluent leaving a dyehouse would

73

Table 1 Composition of 'average' effluent leaving a dyehouse [10]

Component	Amount[a]
Inorganics	17.3 kg m^{-3}
Auxiliaries	2.5 kg m^{-3}
COD	1400 mg l^{-1} oxygen
Suspended solids	280 mg l^{-1}
Colour at 400 nm	0.96 AU
Colour at 450 nm	1.11 AU
Colour at 500 nm	1.37 AU
Colour at 550 nm	1.49 AU
Colour at 600 nm	1.56 AU
Colour at 650 nm	0.87 AU
Colour at 700 nm	0.13 AU

a AU – absorption units in 1 cm cell

contain the components shown in Table 1 [10]. The water quality required at the end of the treatment plant depends on the end-purpose of the water. Clearly, the constraints applied to waters treated for discharge will be different from those applied to water required for recycling within the plant, for which a quality approaching that of softened potable water may be needed. The data in Table 1 also indicate that the major demand on any system designed to allow maximum re-use of the water is the removal not of colour but of the very high inorganic loading, of which the major component is sodium chloride.

The treatment technologies that have been used in attempting to resolve this effluent problem can be grouped into two main categories: those that concentrate the colour, COD, etc. into a concentrate or sludge that then has to be disposed of, and those that actually destroy the colour and other organic components of the effluent. This article will, for each of these groups, outline the technologies from a technical point of view. Economic information will also be considered where it is available in the open literature.

2 Hazards posed by effluents

Before examining individual effluent treatment technologies, some idea is needed as to the potential hazards of the various components discharged. In recent years at least two of the common dye classes (direct and azoic) have been, or are being, considered by European and American authorities as possible carcinogens [11,12]. There appears, however, to be some disagreement as to whether the possible risk lies only with the azoic dyes or includes all dyes that contain an azo group within their molecular structure [11].

Reactive dyes are one of the most common types of dye used and in 1986 the Health and Safety Executive issued publications concerning the possible respiratory irritant and sensitisation effects from handling the dry powders or from aerosols containing these materials [13,14]. In the effluent the concentration of reactive dye components is reduced by the presence of relatively large quantities of inorganic salts and of other organic material. A possible hazard could still exist in aqueous solution and should not be ignored. Besides the dyes themselves, other chemicals or additives (including surfactants) used in either the dyeing or the finishing steps have also been reported to pose health hazards [15–18].

As a result of these known potential problems it would be wise to consider that *any* concentrate or sludge formed from treating dyehouse effluent could well contain hazardous materials. Appropriate technology should be used to dispose of these products, for example, to registered landfills or by incineration.

3 Treatment technologies

There are two possible locations for any technology which could be used to treat the effluent:

— at the dyehouse, either to remove colour and possibly COD or, preferably, to allow partial or full re-use of the water

— at the sewage treatment works, to treat the colour remaining either before the current biological or chemical treatment or as a final polishing step. Either alternative will cause the cost of sewage treatment to increase, which will be reflected in the charges made to the dyehouse.

Ideally the effluent should be at least partially treated on the factory site, but for the short term and until adequate and economic treatment technologies become available the effluent treatment to remove colour is more likely to be at the sewage treatment works. To date the following technologies have all been tried:

— coagulation and/or flocculation, coupled with some form of filtering process and sludge thickening

— membrane technologies such as reverse osmosis, nanofiltration or dialysis

— use of adsorbents such as granular activated carbon, high-surface-area silica, clays, fly ash, synthetic ion-exchange media, natural bioadsorbents (chitin), synthetic bioadsorbents (cellulose ion-exchange media)

— chemical oxidation technologies, including the use of Fenton's reagent with hydrogen peroxide, photocatalysts with ultraviolet irradiation, advanced oxidation processes (UV/peroxide, UV/ozone, peroxide/ozone, peroxide alone, ozone alone or chlorine-based oxidants)

— biochemical oxidation and to a more limited extent bioreduction.

75

Owing to the nature of the effluent to be treated, it is unlikely that a single technology will allow the total on-site treatment of the effluent and enable most of the water to be economically recycled.

4 Concentration technologies

4.1 Coagulation/flocculation

The use of chemicals to generate a precipitate, which either during its formation or as it settles entrains other unwanted species, is a well-established method for water purification. In general terms, the technique requires pH control and the addition of either an inorganic or an organic species that will generate a precipitate. This precipitate is then removed by either flotation, settling, filtration or other physical technique to generate a sludge that is normally further treated to reduce its water content.

Not all dyes are effectively removed by the common materials used to generate the precipitate; for example, alum is unsatisfactory for the removal of colour generated from azoic, reactive, acid and basic dyes, but is good for treating disperse, vat and sulphur dyes [19]. Combinations of various inorganic chemicals have been used to improve colour removal from effluents containing the more common dye types; considerable improvements can be obtained, but at the cost of forming relatively large volumes of sludge [20].

The generation of the precipitates need not be by the addition of chemicals but can also be by electrochemical techniques using sacrificial electrodes [21,22].

The advantages of using the chemical coagulation/flocculation approach are:
– relatively simple equipment, which need not have exceedingly high capital costs
– relatively rapid removal of colour
– significant reduction of COD and, if filtration is used, of suspended solids.

The disadvantages are:
– depending on the chemicals used, considerable volumes of sludge may be generated; these could contain hazardous materials requiring disposal to a registered waste facility
– the chemicals need to be added on a continuous basis
– running costs are relatively high

– carry-over of polyelectrolytes in the liquor can have detrimental effects on sewage treatment, especially on denitrification.

5 Membrane technologies

The use of various forms of membrane to clean effluent from textile dyers and finishers is well documented in the open literature, with virtually every membrane technique having been tried [23–32].

5.1 Reverse osmosis

The most frequently tried approach uses reverse osmosis. Here the effluent is forced under moderate pressure across a semipermeable membrane to form a purified permeate and a concentrate [23,26,28–30]. The process can remove up to approximately 98% of the impurities in the water with a relative molecular mass (r.m.m.) in excess of 100. Even with this high degree of efficiency, the permeate may still contain a residual salt concentration that is too high for recycling to the front end of the factory. The permeate usually accounts for up to 80% of the total flow into the unit, with the retentate accounting for the remainder. The concentration of impurities in the retentate will be approximately four times that in the original effluent, and it will still require treatment by some other process before disposal.

The membranes in these units have to be cleaned on a regular basis and may be attacked by the dye materials or other constituents of the effluent; this may change their surface characteristic, resulting in either a poorer-quality permeate or premature membrane failure.

Reverse osmosis units in general are associated with high capital costs and relatively high running costs. For this particular application the use of reverse osmosis on its own has the following disadvantages:
– high capital costs
– at least 20% of the total effluent is not treated
– the concentrate contains virtually all the impurities from the factory effluent
– the concentrate has to be treated by some other technology
– the purified stream may still contain too high a level of impurities for recycling.

5.2 Nanofiltration

A more recent development is nanofiltration, where the membrane effectively acts as a molecular filter retaining material with an r.m.m. greater than about 200 [24]. This technique requires the effluent to be circulated significantly more rapidly than for reverse osmosis, but can achieve concentration factors in excess of ten. Again, the concentrate generated contains virtually all the organic impurities and some of the inorganic impurities from the effluent, and requires treatment by an alternative technology. As with reverse osmosis, the membranes need frequent cleaning, and again there is a possibility that dyes or other components of the effluent could permanently attach to or react with plastic membranes. While both ceramic and stainless steel nanofiltration membranes are marketed, their cost is currently too high to allow their use in this application.

Nanofiltration units have very high capital costs, and higher running costs than reverse osmosis units of comparable size. They can, however, generate a smaller volume of concentrate than a reverse osmosis unit but the permeate will have a much higher concentration of inorganic salts. The major disadvantages of nanofiltration, as stand-alone units, are similar to those of reverse osmosis, namely:

– high capital costs
– formation of a concentrate that can be up to 10% of the total volume treated
– the concentrate contains virtually all of the organic components from the effluent treated, but only part of the inorganic salts
– the concentrate needs to be treated by an alternative technology
– the permeate may still contain an unacceptably high concentration of inorganic salts for water recycling and therefore may need additional treatment by reverse osmosis.

5.3 Ultrafiltration and microfiltration

Ultrafiltration and microfiltration have both been applied to the treatment of effluent from textile processing. Their main application has been to reduce suspended solids and higher-r.m.m. organic materials with particle sizes in the region of 0.02 mm (ultrafiltration), or greater, to form a sludge for disposal [27,30,31]. Neither of these techniques has any major effect on the concentration of inorganic salts, nor will they reduce the colour unless they are adsorbed on to other material that is removed. Since these techniques do remove submicron particulate they will have an effect on the apparent colour, especially at the lower wavelength, that is really due to light scattering. As stand-alone techniques these two approaches are suitable only for the reduction of COD and suspended solids, but they will prove very useful as part of a multi-technique effluent treatment unit.

5.4 Dialysis or continuous deionisation

An alternative membrane technique uses an electric field across a dialysis unit, with the inner cell separated from the two outer cells by ion-exchange membranes (Figure 1) [32]. The process generates two concentrate streams, one containing all the anions and the other containing only the cations. Since virtually all the cations are those of sodium, this technique in theory could be used to generate a brine solution which could be recycled within the process.

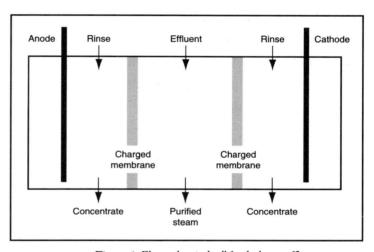

Figure 1 *Electrochemical cell for dyehouse effluent treatment*

Since the original Russian paper appeared in the early 1980s, no further work has been published concerning this technique for this application. A similar unit has, however, been commercialised for the purification of water for laboratory and other industrial applications. The advantages of this approach are:

– the purified stream could be suitable for recycling
– the cation concentrate could be re-used either to regenerate a conventional water softener or, if sufficiently pure, as either caustic or sodium carbonate in the dyeing process.

The disadvantages are similar to those of reverse osmosis or nanofiltration, namely:

– a portion of the total effluent is effectively not treated

– the concentrates will contain virtually all the impurities from the factory effluent

– at least one of the concentrates has to be treated by some other technology

– the purified stream may still contain too high a level of impurities for recycling

– organic material present could foul the membranes

– nonionic species are not removed from the effluent

– capital and operational costs are unknown for this application.

6 Adsorbents

The use of any adsorbent, whether ion-exchanger, activated carbon or high-surface-area inorganic material, for removing species from a liquid stream depends on the equilibrium between the adsorbed and the free species. For the process to work at its maximum efficiency the concentration of impurities in the feed stream should be as constant as possible in order to avoid the release of adsorbed material back into the stream if the concentration falls. In multicomponent mixtures not all of the species are bound to the adsorbent to the same degree, with the result that strongly bound species may displace more weakly bound components.

The range of adsorbents described in the literature for this application covers the range of activated carbons [19,33–36], high-surface-area inorganic materials [35–37], synthetic ion-exchange resins [34,38–46], cellulose-based adsorbents such as chitin (poly-N-acetyl-glucosamine) [33,47–60] and synthetic cellulose and other fibre-based bioadsorbents [61–63].

6.1 Activated carbon

Activated carbon is reasonably effective at removing organic components from aqueous streams but once it is fully loaded it needs either regeneration or disposal [19,33–36]. The effective cost of the high-temperature regeneration process, including the cost of replacement carbon to make up for that lost as carbon dioxide, makes regeneration unattractive to the small user, as it often requires the carbon to be removed from site and shipped to a specialised facility. Disposal requires great care, as any organic component adsorbed on the activated carbon can be desorbed or leached out over a period of time, and if the disposal site is not adequately protected can cause pollution in other watercourses or aquifers. Activated carbon does not remove the inorganic components from the effluent stream and therefore would not allow, if used on its own, the recycling of the water to the front of the factory. It is therefore not surprising that this process has not gained favour for large-scale applications. It may, however, be used economically as a final polish to a multistage effluent plant.

6.2 Inorganic adsorbents

The use of inorganic adsorbents, such as high-surface-area silica, cinder ash, clays and so forth, has been tried for a range of dyes [37,64–66]. Their effectiveness depends on the types of dye in the effluent stream or, more particularly, with the relative charge on the dye molecule or fragment.

Silica was found to be reasonably effective for treating effluents containing basic dyes (positively charged) but virtually inactive for effluents containing dyes that are effectively negatively charged, such as reactive dyes. Again the process has little effect on the major inorganic charge of the effluent, and additional treatment would be required if the treated effluent was to be recycled. Regeneration of the inorganic systems is in theory simpler and can be achieved by several techniques, including allowing a biomass to grow in the bed, effectively using the organic pollution load as a food supply [65].

The use of cinder ash appears to be a cost-effective solution where a readily available supply of the ash is available locally and where the effluent does not contain reactive dyes. Two such units have been developed by the Textile Institute of China and have been operating for some years [65].

6.3 Ion-exchange resins [34,38–46]

Since most dyes are chemically either anionic (reactive dyes, acid dyes, direct dyes, etc.) or cationic (basic dyes), they could in theory be removed on ion-exchange resins. The extremely large inorganic load of the effluent effectively precludes the use of conventional resins, which are selective for small anions and cations. Work has been reported on the use for this application of macroporous resins, which remove the larger organic ions rather than the smaller inorganic ions. At least one pilot study has demonstrated that the process is feasible providing an organic solvent such as methanol is used during the regeneration step [46]. The methanol is recovered and recycled to the process leaving behind a solid, mainly organic mass which has to be disposed of to either incineration or other hazardous waste site. If the effluent initially contained reactive dyes this solid residue would need to be treated with extreme care due to the known respiratory sensitisation and irritant effects of these materials [13,14].

6.4 Bioadsorbents

Bioadsorbents are naturally occurring polymers (or their synthetic derivatives) which are biodegradable and which have structures that either adsorb species within them or under the operating conditions act as ion-exchangers. Most of the published work refers to the use of purified chitin or (more precisely) chitin containing a small amount of chitosan as an impurity, rather than the raw waste materials from the shellfish industry from which both chitin and chitosan are extracted [67]. This particular material was first studied for this application in the 1950s [57–59]. Since the early 1980s interest in its uses and properties has increased and a considerable amount of information is now available.

McKay *et al.* have reported detailed studies concerning the uptake of various pure dyes and the time required for the adsorbed and free dye to come to equilibrium in the presence of chitin [47–51,55,56], and have mentioned its chemical instability [49]. They have shown that the material is suited only to dyes that are anionic, or weakly anionic, in nature and that this material under the test conditions had relatively low capacities for these dyes (fractions of a milliequivalent per gram of chitin) and required something in the region of five days to reach equilibrium. The principal component for dye removal is clearly chitosan, with a direct relationship between chitosan concentration and quantity of dye removed [52–54]. The effect of pH on dye loading has also been examined; low-pH conditions are required, suggesting that the chitosan was acting as a weak-base anion-exchanger; this confirms the original work by Giles *et al.* [57–59]. The chemical instability of chitin under these conditions has been studied, and attempts have been made to crosslink the polymeric structure to increase the resistance to chemical attack. This resulted in further reducing the capacity for the dyes.

Preliminary results from batch tests under aerobic conditions are currently under way to establish the type and degree of chemical attack under acidic conditions. These have shown that in the presence of 5 mmol l^{-1} hydrochloric acid, chitin is chemically attacked to produce carboxylic acids (acetic acid and formic acid), as well as significant quantities of nitrite (>5 mg l^{-1} as NO_2^-) and dissolved total organic carbon (TOC) [68]. Even in the presence of ultrapure water low levels of nitrite can be detected. The nitrite levels are not dependent on initial nitrate concentration but appear to be affected by pH, chloride concentrations and other factors as yet not identified. These results confirm that chitin is chemically unsuitable for use under these operational conditions (pH 3–4) and would have limited bed life of approximately a few months.

The effects of various other dyehouse effluent components on dye removal using chitin have been studied, and have shown that both dye mixtures and surfactants can reduce significantly the amount of dye removed by a given mass of chitin [52]. If the principal binding mode of the dyes to chitin is via a weak-base anion-exchanger, as suggested above, then the very high levels of inorganic salts found in dyehouse effluents would also be expected to have a negative effect on the dye-binding capacity of chitin.

The effect of temperature on dye removal has also been examined and found to be dye-dependent [55,56]. For small-molecule dyes which are readily adsorbed, increasing the temperature decreases the amount of dye removed; on the other hand, for very bulky dyes which are only poorly adsorbed an increase in temperature causes an increase in adsorption.

The results for chitin can be summarised as follows:
- low capital cost for a once-through process
- a high cost for purified adsorbent (£10–20 per kg), but low cost for raw shellfish waste (US$80 per tonne)
- since the effluent needs to be at low pH, there is a risk of toxic or harmful gaseous emissions formed by interaction of the acid with other effluent components such as sulphides or chlorinating agents [10]; monitoring and fail-safe systems would be required for these materials.

6.5 Synthetic cellulose and other fibre-based bioadsorbents

In the early 1980s the Institut Textile de France studied the use of cellulose-based ion-exchange materials for the treatment of dyehouse effluents. Because of the synthetic nature of these materials, they could be tailored to have greater chemical resistance; they are said not to produce nitrite under operational conditions [62]. They are reported to have significantly higher capacity than chitin (milliequivalents per gram) and do not need to be loaded under acid conditions. Both anion- and cation-exchangers have been made, so that the whole range of dissolved dyes can be treated. At least one UK manufacturer of these materials has undertaken successful preliminary studies for effluents containing wastes from the use of reactive dyes [62]. French workers are now developing more inert fibrous backbones for these materials [63]. The properties of cellulose-based bioadsorbents can be summarised as follows:

– less susceptible to chemical attack than are chitin-based materials
– both cationic and anionic varieties available
– higher capacity than natural bioadsorbents
– can be regenerated a significant number of times
– have been used by Institut Textile de France for treating the effluent from dyehouses using both reactive and basic dyes in pilot field study
– can be tailor-made to suit a given application.

If this group of synthetic cellulose-based adsorbents really prove to be as good as the preliminary work has suggested, they could be extremely useful in removing colour for those sites where there is insufficient space to install an actual effluent treatment facility but where there is a requirement for off-site regeneration and treatment of the colour concentrate produced. The economics of their use in the UK are as yet unknown.

7 Destruction technologies

In the light of the proposed EU legislation looking at an integrated approach to waste treatment [69], it is likely that, in general, destruction technologies will gain favour at the expense of those technologies that just transfer the pollution charge from the liquid phase to the solid phase for disposal or form a liquid concentrate for further treatment.

The best-established destructive treatment technologies for dyehouse effluent treatment are based on biological oxidation (aerobic digestion) or reduction (anaerobic digestion), and plants of this type are operating in the UK. These plants are in general only suitable for sites where a reasonable amount of space is available, and are not capable of effectively treating all types of dye.

7.1 Aerobic digestion

Conventional aerobic biological digestion is the basis of most sewage treatment works. It uses a biomass to convert the incoming BOD charge into carbon dioxide and a sludge, which is frequently transferred to an anaerobic digester for further treatment before finally being disposed of for various applications including spreading on fields as fertiliser. One of the major disadvantages of an aerobic digester is that it generates a considerable amount of sludge, which may contain adsorbed species that have passed through the treatment without being retained. The liquor resulting from aerobic digestion may still contain significant colour and may require polishing to achieve the proposed NRA consent limits. At least one company is claiming to improve the rate of digestion by using pure oxygen, allowing a given plant to process a greater charge and also to a final liquor of better quality [70]. The same results may be obtainable by using so-called 'deep shaft' technology, where compressed air is effectively used to increase oxygen concentrations and biomass respiration efficiently.

7.2 Anaerobic digestion

Essentially anaerobic biodegradation converts the organic pollutants via a series of metabolic reactions to end-products that are mainly gaseous (principally methane and carbon dioxide) as opposed to aerobic degradation, which forms mainly solid end-products. The methane produced can be used to generate thermal energy on site for use within the process. The conversion of the organic pollutant charge depends on the presence of several species of bacteria, each interdependent on the others. Since not all these bacteria grow at the same rate, control over the rate at which the pollutant charge enters the system is necessary and balancing tanks are required in order to remove sudden changes in charge. One author has reported that the use of two reactors to undertake the treatment can allow more control and achieve lower final COD levels in the liquor [71].

Anaerobic systems are very expensive to install but do not require the large energy requirements of an aerobic system and generate only a relatively small volume of sludge.

8 Chemical oxidation technologies

By suitable control of reaction conditions and methodologies, two scenarios can be obtained:

partial oxidation of the organic charge to destroy the conjugated bonding system in the chromophore, thus destroying the colour, and *total oxidation* (or mineralisation) of the organic charge to form carbon dioxide and inorganic ions. The survey of dyehouse effluents into one sewage treatment works suggested that an average COD load would be in the region of 100 mg l^{-1} of oxygen [9]. This would translate as TOC in the region of 2000 mg l^{-1} as carbon and would require 8000 mg l^{-1} ozone or approximately 19 ml of 30% hydrogen peroxide per litre. Assuming an oxidant cost of approximately £500 per tonne for peroxide and 100% utilising efficiency, then peroxide costs alone would be in the region of £9.50 m^{-3}. Costs for ozone would be expected to be at least 25% higher. These figures show clearly that total chemical oxidation using either of these chemicals is not economically viable, although various authors have proposed the use of either ozone [72], ozone in conjunction with UV [73–75], Fenton's reagent in conjunction with peroxide [76,77], UV in conjunction with peroxide [78] or peroxide alone [79,80]. Systems using oxygen or air may be economically viable but as yet no low-temperature systems exist, although a high-temperature system has been installed in the Far East [81].

The problem associated with partial oxidation lies in the unknown nature of the products formed and the risk that some of these may be even more harmful to the environment than the initial components of the effluent. However, the amount of oxidant required would be only a small fraction of that required for total oxidation, and the oxidants need not necessarily be ozone or peroxide but could be based on any of the chemical oxidants currently available including hypochlorite [73], persulphate (often used as oxidant in total organic carbon analysers) and others. Hypochlorite or chlorine dioxide treatment of effluents at pH values less than 8 could well result in the formation of chlorinated organic species, which themselves can pose an environmental problem, and which if formed would need to be either removed or destroyed. Alternatively physical methods of oxidation which generate singlet oxygen and hydroxyl radicals could be used, such as titanium dioxide photocatalyst in conjunction with UV irradiation [82,83], oxygen in conjunction with hard UV [84], or ultrasound [85]. For the processes using UV radiation, the energy efficiency of the lamps is of prime importance and is currently best in the region of 25%. In photocatalytic process the efficiency of conversion of UV into effective radicals is less than 1% and this process is thus very energy-intensive.

Nearly all chemical oxidation processes take place via radical mechanisms. Once the reaction has started, therefore, control over the reaction products will be difficult even in laboratory systems, and virtually impossible using a variable feedstock such as an industrial effluent. It is thus extremely important that the reaction products from these partial oxidative systems are studied before large-scale commercial installations are built. The hydroxyl radical oxidation of monochlorinated species is known to produce small amounts of polychlorinated species which are eventually destroyed [85,86].

The advantages of chemical oxidation include:

– partial oxidation is economical, as it involves both low capital and running costs
– total oxidation generates only carbon dioxide and inorganic ions
– the process can be made extremely rapid, thus minimising equipment size and cost.

The disadvantages are:

– total oxidation with either peroxide or ozone is extremely expensive
– chemical oxidation only reduces carbon loading of the effluent, and can increase the concentration of inorganic material in solution
– energy costs for photocatalytic systems are high
– the economics for total oxidation using oxygen and hard UV are unknown
– partial oxidation could generate species that may be more harmful to the environment than those in the original effluent.

9 Discussion

In order to treat effluents from dyehouses and allow the water to be recovered and re-used, a combination of technologies will be required.

Reverse osmosis and dialysis or continuous deionisation appear to be the only currently available technologies that can significantly reduce the inorganic loading of the effluent. Even these technologies, however, might require multiple treatment stages to bring the salt level down to an acceptable level. They have high capital costs.

The concentrate or concentrates from the membrane systems would need to be treated, ideally by a destructive technology such as biological system or total chemical oxidation. Again these systems have a high capital cost and/or high running costs.

The membrane systems would also need to incorporate some form of pretreatment to remove all suspended solids and if necessary to remove any components that could prematurely destroy the membranes, such as free chlorine or any component that could react with the membrane surface. Some process may always be needed to deal with any solids generated either during the treatment of the concentrate or concentrates from the membranes or during the pretreatment steps.

At current water and sewage costs, it is thus highly unlikely that a treatment plant that allowed most of the water to be recycled could be economic. The alternatives will be to treat the effluent on site to eliminate colour (and possibly reduce COD and suspended solids) or to remove the colour at the sewage treatment works. Clearly any on-site treatment will need to form part of a long-term integrated plan which would include the eventual recycling of process water. These short-term solutions are more likely to be based on the established technologies which are not capital-intensive, or on newer technologies such as partial chemical oxidation, providing research has confirmed that the partially oxidised products are not more harmful to the environment than those originally discharged.

10 Conclusions

The principal conclusion to be drawn is that the technologies that would allow the process water to be recycled are associated with high capital costs, and are therefore not very attractive to the textile dyers and finishers. The lower-cost technologies do not allow the effluent to be treated to a quality that would allow process water recycling but, providing adequate safeguards are put in place, could be suitable for a short-term solution and form part of a longer-term integrated approach that would eventually allow the process water to be re-used.

It is also clear that independent research needs to be carried out to identify possible hazards associated with any of the processes mentioned in this article but especially processes such as partial oxidation including the use of hypochlorite, chlorine dioxide and ozone. Ideally, this work should be funded by the technology suppliers but supervised by one of the relevant trade associations, and be started immediately.

Finally, it is clear from the literature that some processes are unsuitable for this application on technical grounds, because the materials they use are attacked chemically by components of the effluent or because of the need to use conditions which are known to attack part of the process.

11 References

1. T G Southern, Proc. Knitting Industries' Federation meeting, Hinckley, Leicestershire (Aug 1993).
2. P Cooper, *J.S.D.C.*, **109** (1993) 97.
3. W S Hickman, *J.S.D.C.*, **109** (1993) 32.
4. B G Hazel, *J.S.D.C.*, **107** (1991) 392.
5. I G Laing, *Rev. Prog. Coloration*, **21** (1991) 56.
6. B D'Arcy, *J.S.D.C.*, **107** (1991) 387.
7   B G Hazel, *J.S.D.C.*, **107** (1991) 204.
8. F A N van Baardijk, *Ind. Env.*, **13** (1990) 3.
9. K Schliefe, *Melliand Textilber.*, **68** (1987) 217.
10. S F Watts *et al.*, to be published.
11. *Chem. Br.*, **26** (1990) 820.
12 *Health and Safety at Work*, **12** (1990) 7.
13. HSE leaflet Ind. (G) 82L.
14. HSE leaflet on reactive dyes 765/4.
15. *Eur. Chem. News*, **57** (1992) 26.
16. *Hazards*, **11** (1986) 6-7.
17. *Guidelines for the safe storage and handling of non-dyestuff chemicals in textile finishing* (HSC: Nov 1985).
18. K Kawai *et al.*, *Dermatitis*, **28** (1993) 117.
19. *Developments in chemistry and technology of organic dyes, critical reports on applied chemistry*, Vol. 7, Ed. J Griffiths (London: Blackwell Scientific, 1984).
20. T Panswad and S Wongchaisuwan, *Water Sci. Technol.*, **18** (1986) 139.
21. Zhu Youchun *et al.*, *Water Treatment*, **6** (1991) 227.
22. C Heinitz, *Wasser und Boden*, **35** (1991) 30.
23. J D Nirmal *et al.*, *Separation Sci. Technol.* (New York), **27** (1992) 2083.
24. *Membrane Technol.*, (20) (1991) 6.
25. J C Watters, E Biagtan and O Senler, *Separation Sci. Technol.* (New York), **26** (1991) 1295.
26. V Calabro *et al.*, *Desalination*, **78** (1990) 257.
27. K Majewska-Nowak, T Winnicki and J Wisniewski, *Desalination*, **71** (1989) 127.
28. R L Riley and P A Case, *Desalination*, **36** (1981) 207.
29. E Staude, *Chemie Ing. Technik*, **45** (1973) 12.
30. C A Buckley, *Water Sci. Technol.*, **25** (1992) 203.
31. R B Townsend *et al.*, *Water SA*, **18** (1992) 81.
32. M L Panomarev and V D Grebenyul, *Electrokhimiya*, **12** (1976) 823.
33. G Mackay, J C Kelly and I F McConvey, *Adsorption Sci. Technol.*, **8** (1991) 13.
34. T Saito, K Hagiwaro and P Kusano, *Mizu Shori Gijutsu*, **26** (1985) 469.
35. *Colourage Annual*, (1989/90) 83.

36. G McKay *et al.*, *Colourage*, **27** (1980) 3.

37. G McKay, M S Otterburn and A G Sweeney, *Water Res.*, **14** (1980) 15.

38. D C Kennedy, *Chemical Eng. Prog. Symp. Series* (152) (1976) 71.

39. D C Kennedy *et al.*, *Amer. Dyestuff Rep.*, **63** (1974) 8.

40. B Zhou *et al.*, *Zhonghua Yufang Yixue Zazhi*, **22** (1988) 338 (*Chem. Abs.*, 11: 139 817).

41. F Cognigni and E Sciolla, Eur Pat Appl. EP237 101 (1987).

42. I M Mukhamedou, V K Kryuchkoic and A Z Zgibnava, *Uzh. Khim. Zhur.*, (1986) 48.

43. M Nakau and Y Sato, *Mizu Shori Gijutsu*, **23** (1982) 222 (*Chem Abs.*, 98: 166 426).

44. A V Shvangiradze *et al.*, *Izv. Akad. Gruz. SSR Ser. Khim.*, **4** (1978) 271 (*Chem Abs.*, 90: 174 103).

45. S L Rock and B W Stevens,. *Text. Chem. Colorist*, **7** (1975) 169.

46. A Maggiola and S J Sayles, US NTIS PB-273362 (*Chem Abs.*, 88: 176 701).

47. G McKay, M S Otterburn and A G Sweeney, *Water Res.*, **15** (1981) 327.

48. G McKay *et al.*, *J. Colloid Interface Sci.*, **80** (1981) 323.

49. G McKay, H S Blair and J Gardner, *J. Colloid Interface Sci.*, **95** (1983) 108.

50. G McKay, H S Blair and J Gardner, *J. Appl. Polymer Sci.*, **29** (1984) 1499.

51. G McKay, J C Kelly and I F McConvey, *Adsorption Sci. Technol.*, **8** (1991) 13.

52. Y Shimizu *et al.*, *Mizu Shori Gijutsu*, **29** (1988) 43.

53. Y Inoue *et al.*, *Shiga-Kenritsu Tanki Daigaku Gakajutsu Zasshi*, **27** (1985) 1.

54. Y Inoue *et al.*, *Shiga-Kenritsu Tanki Daigaku Gakajutsu Zasshi*, **25** (1984) 6.

55. G McKay, H S Blair and J Gardner, *J. Appl. Polymer Sci.*, **27** (1982) 3043.

56. G McKay, H S Blair and J Gardner, *J. Appl. Polymer Sci.*, **27** (1982) 4251.

57. C H Giles *et al.*, *J.S.D.C.*, **74** (1958) 647.

58. C H Giles, A S A Hassan and R V R Subramamian, *J.S.D.C.*, **74** (1958) 682.

59. C H Giles, A S A Hassan and R V R Subramamian, *J.S.D.C.*, **74** (1958) 846.

60. D Knorr, *Food Technol.*, **45** (1991) 114, 116, 122.

61. D Wattiez and E Marechal in *Polymers, amines and ammonium salts, Int. Symp.*, Ed. E J Goethals (Oxford: Pergamon, 1980) 357.

62. R Highton, Whatman Ltd, personal communication.

63. Institut Textile de France, personal communication.

64. G McKay, M S Otterburn and A G Sweeney, *Water Res.*, **14** (1980) 21.

65. Q Zhao and G Li, *Water Sci. Technol.*, **24** (1991) 215.

66. *Filtration Separation*, **29** (1992) 485, 487.

67. *Chem. Eng. News*, (Summer 1993).

68. S F Watts *et al.*, personal communication.

69. *Europe Env.*, (417) (Sept 1993).

70. *BOC Newsline*, BOC G7172/LGM/1093/NT/6C.

71. P Ditchfield, *Trends in Biotechnol.*, **4** (1986) 309.

72. F U Gaehr, Dissertation, University of Stuttgart (1992).

73. S Patel, Project dissertation, Oxford Brookes University (1994).

74. S V Prejs and E K Sijrde, *Tr. Tallin. Politekh. Inst.*, **658** (1988) 16.

75. O Leitzke, *Brauwelt Int.*, (1993) 422.

76. W G Kuo, *Water Res.*, **26** (1992) 881.

77. K H Gregor, *Melliand Textilber.*, English Edn (1990) E435.

78. M Pitroff and K H Gregor, *Melliand Textilber.*, English Edn (1992) E224.

79. U Gruntz and A Wyss, *Z. Wasser- und Abwasser-Forschung*, **23** (1990) 58.

80. H Debellefontaine *et al.*, *Information Chimie*, (323) (Dec 1990).

81.   J Haggin, *Chem. Eng. News*, (4 Feb 1991).

82.   A Mills, R H Davies and D Worsley, *Chem. Soc. Rev.*, (1993) 417.

83.   R J Davies *et al.*, *Water Env. Res.*, **66** (1994) 50.

84.   Hanovia UK, personal communication.

85.   N Serpone *et al.*, *J. Phys. Chem.*, **98** (1994) 2634.

86.   T G Southern, unpublished work.

CHAPTER SEVEN

The treatment of dyehouse effluent at Stevensons Fashion Dyers

John Scotney

1 Introduction

The Stevenson family originally set up a dyeworks on the banks of the River Amber in Derbyshire at the turn of the century, mainly to carry out the dyeing and cleaning of clothing. Over the years, the operation has increased in scale so that use is now made of most of the 7 ha site shown in Figure 1. The effluent treatment plant is the series of tanks running parallel to the River Amber and the railway line in the top left of the aerial photograph; it occupies an area of 0.2 ha.

Figure 1 *Stevensons Fashion Dyers: the dyeworks*

The basic process used is the batchwise dyeing and finishing of, mainly, knitwear such as jumpers, socks and tights, on a commission basis. Most of the merchandise reaches the ultimate consumer via the major high street retailers. Production is now at a level of over 3 million units per week, involving about 180 tonnes of fabric and the weekly consumption of water can approach 50 000 m^3. This water is obtained from several sources: a borehole on the site, the River Amber and from Severn Trent Water. All the dyehouse effluent must be treated before it can be put back into the Amber.

The weight of a dye lot is usually in the range 50–200 kg and there are over a hundred dyeing machines of various designs, operating round the clock apart from the weekend. Processes carried out in these machines range from acid chlorination on wool to alkaline bleaching of cotton and from boiling dyebaths to cold rinses, with virtually every class of dye, including pigments, applied to every substrate and a variety of blends. Typically, the dyeworks effluent contains a mixture of waxes found naturally in fibres such as wool and cotton, oils used to spin the yarn, detergents, auxiliary chemicals used to control the dyeing process and unfixed dyes.

2 Biological treatment plant

Historically, the effluent was treated by running it through a series of channels to get mixing and then contacted with aluminoferric to adsorb some of the impurities. The insoluble floc was allowed to settle in a lagoon and the treated effluent was discharged into the River Amber. The settled sludge was pumped out and dewatered on a filter press. Over the years, this process proved inadequate to cope with the increasing pollution load and volume, and pilot plant trials were carried out in the mid-1960s to establish what technology would be best suited to treat the dyehouse effluent to Royal Commission standards, i.e. BOD/suspended solids 20/30 ppm. Among the systems looked at were biological treatments, using either activated sludge tanks or percolating filters, and chemical flocculation followed by settlement or flotation. At this time the major fibres being dyed were wool, nylon and acrylic, and thus all the processes used were characterised by high levels of dye exhaustion.

Apart from the dyehouse effluent, two other waste products being produced on site needed to be considered.

(a) Incoming river water and borehole water was treated by the lime-soda process to clarify and soften the water for process use. This produced a sludge of calcium carbonate and magnesium hydroxide. It was desirable to feed it into the dyehouse effluent stream at the head of the treatment plant.

(b) The coal-fired boilers discharged finely powdered ash as the waste product of combustion. This was available as a filter aid to assist in the dewatering of any sludges produced during the effluent treatment.

It was finally decided to use a modified activated sludge treatment, the contact stabilisation process, and this plant was installed and commissioned by Permutit in the late 1960s. It is still in use. Figure 2 is a schematic summary of the various stages of the effluent plant; the Permutit installation is shown as pre-1986, above the broken line. All the waste water from the dyehouse is led to the site of the original treatment plant, which is now used as a collecting sump and contains coarse screens to remove larger items of waste. Loose fibre and lint is filtered out on a fine rotary screen before the raw effluent is pumped into a large balancing

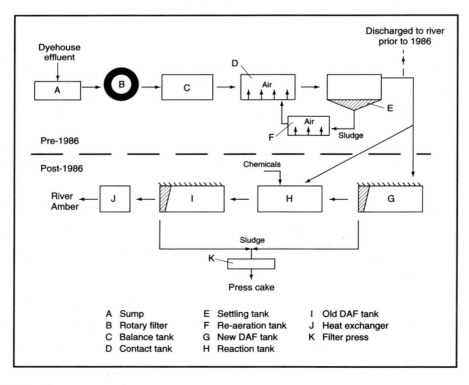

Figure 2 *The effluent treatment plant*

tank. Holding about 1200 m³, equivalent to 3–4 hours' production, this tank evens out the composition, gives an average temperature of 38–40 °C and protects the living organisms in the subsequent biological tanks from any shocks. The original design used air injected into the base of the tank to mix the contents. It has been found since that this aeration process actually reduces the total oxygen demand of the raw effluent by some 15%. The effluent contains quantities of sulphite from the wool shrink-resist treatments, and this is best removed by simple chemical oxidation so that it does not compete with the biomass for the available oxygen in the next stage of the treatment.

Variable-speed pumps transfer the mixed effluent at as constant a rate as is possible into two activated sludge tanks running in parallel. Here, it is contacted with biomass which absorbs the impurities from the water. In the presence of air, injected through diffusers, the bacteria use the waste material as food, proceeding to grow and multiply to produce an ever-increasing amount of biomass. After a contact time of three hours at a temperature of 40 °C, the mixed liquor is passed to a clarifier; here the biomass is allowed to settle under gravity and the clear purified water flows off the top of the tank. The settled sludge is pumped into a re-aeration tank, where the biomass completes the digestion of absorbed nutrients before it is contacted with fresh effluent. Although Figure 2 shows three completely

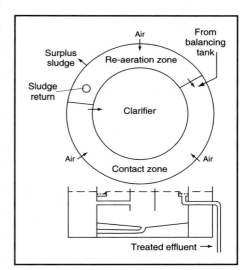

Figure 3 Biological treatment tank

separate tanks to help to explain the contact stabilisation process, in practice large concentric circular tanks are used with the clarifier in the centre and the contact and re-aeration zones arranged as annular zones round it (Figure 3). This gives a very compact configuration; the tanks are 17 m across and 5 m deep.

The concentration of mixed liquor suspended solids in the contact zone is 8000 to 10 000 ppm, and that in the re-aeration zone 14 000 to 18 000 ppm. These figures are higher than those usually found in activated sludge plants; in this installation, however, the proportion of organic biomass material is only about 25% and the balance is made up of the inorganic sludge from the river water treatment plant. Although provision was made in the original plant to dose ammonium phosphate as an additional nutrient for the biomass, it has not been necessary to use this facility because there is already sufficient nitrogen and phosphorus from the dyeing chemicals in the raw effluent.

The biological process results in the conversion of waste material in the effluent into increasing amounts of biomass, which at the Stevensons installation is augmented by the insoluble sludge from the river water treatment. To maintain the concentration of solids in the contact and re-aeration zones at the optimum levels, slurry is drawn from the re-aeration zone and passed to another settling tank for disposal. Originally this concentrated slurry was treated with fly ash and polymer and then filtered off on a rotary vacuum filter for disposal as a 20% solids cake to landfill. By the mid-1970s the coal-fired boilers had been replaced by boilers burning gas/oil; the loss of the fly ash made the vacuum filtration unit much more difficult to operate and eventually disposal was carried out by tankering out the thickened sludge for spraying on to agricultural land as a low-grade fertiliser.

3 Tertiary treatment for colour removal

This was the effluent plant up to 1986, with the treated water being discharged to the River Amber. It was designed to handle 225 m^3 per hour but over the ensuing years production on the site increased, so that now it is required to deal with peak flows of 450 m^3 per hour. Although the biological oxidation stage could cope with these surges, the hydraulic load on the clarifiers proved too much and the effluent quality was not consistently good for suspended solids. Circumstances at Ambergate also changed in the early 1980s, when more and more cotton was being dyed with reactive dyes. Most of the dye forms a strong chemical link with the cotton, but a proportion reacts with the water in the dyebath to become inactive and this goes into the effluent. The hydrolysed dye is not biodegraded in the activated sludge plant and this results in coloured water being discharged into the river. Although the relatively low concentration of dye in the river is not toxic, it did cause an aesthetic problem for Severn Trent Water, who operate a water abstraction plant at Little Eaton, eight miles downstream from Stevensons.

In 1986, a major investment in additional plant was made to remove this dye and to reduce the suspended solids; it is shown schematically in Figure 2, below the dotted line. A specially selected cationic polymer is dosed into the coloured water, reacting with the anionic dyes to form an insoluble product, and the coloured particles are treated with a flocculant in the presence of any suspended solids to produce large flocs in a reaction tank. These are separated from the clean liquid by dissolved air flotation (DAF), a process in which fine bubbles of air are released in a tank containing the suspension, to lift them to the surface to form a stable layer. The advantages of DAF over traditional gravity settling tanks are that it is more compact and cheaper to install and it gives a much more positive separation of solids from the water; on the other hand, it uses a lot of energy to recycle up to 20% of the flow at a pressure of 590 kPa to provide the air for flotation.

At this stage, the water is too warm (40 °C) to be discharged into the river and it is passed through a heat exchanger to preheat the incoming water used in the factory. Apart from conserving energy, the warmer river water is treated more effectively in the lime-soda process. A general view of the new plant is shown in Figure 4.

The solids from the DAF tanks are combined with the surplus sludge from the biological tanks and until very recently the combined solids were disposed of by injection into agricultural land as a fertiliser and soil conditioner.

Figure 4 Stevensons Fashion Dyers: the new effluent treatment plant

4 Recent improvements

Two further modifications have been made in the last two years. Firstly, increasing production,

Figure 5 *The new dissolved air flotation tank*

Figure 6 *The recently installed filter press*

particularly the dyeing of cotton, has led to a heavier load on the colour removal plant. An additional DAF tank has been installed to carry out a preliminary separation of suspended solids prior to the flocculation process carried out in the reaction tank (Figure 5).

Secondly, the uncertainty of being able to dispose of sludge to land, and the increasing cost of this route, has very recently led to a decision to install a filter press (Figure 6). As a result, the quantity of waste material produced each week has been reduced from 250 tonnes of slurry to two skips of press cake. Despite the absence of fly ash, used on the original Permutit plant as a filter aid, the traditional plate press (operating at 620 kPa and with polymer dosing to get flocculation) has no difficulty in producing a cake containing more than 40% solids.

5 Consent limits and performance

The general performance of the plant is summarised in Table 1, which shows the relevant parameters for the effluent before and after treatment and compares them with the consent limits. It will be noted that the limits for BOD/suspended solids (15/25 ppm) are lower than the original Royal Commission values of 20/30; this is because the volume consented in 1987 was

Table 1 Consent limits and performance at Stevensons

Parameter	Raw effluent	Treated effluent	Consent limit
Suspended solids/ppm	300–700	10–20	25
BOD/ppm	80–160	5–12	15
COD/ppm	500–800	80–100	–
Ammoniacal nitrogen/ppm	1–5	1–5	5
pH	8	8	5–9
Temperature/°C	40	28–30	30
Absorbance at 500 nm	–	0.01–0.03	0.05
Metals/$\mu g\ l^{-1}$		100–600	1000
PCP/$\mu g\ l^{-1}$	<0.5	<0.5	–

increased from 6750 to 9000 m^3 per day. The pH value is the natural pH of the effluent and is due to the mixed product range being dyed and treated under both acidic and alkaline conditions, and the buffering effect of the sludge from the lime-soda process.

The full colour consent is a list of optical density values through the spectrum from 400 to 700 nm. A peak at 500 nm denotes a red coloration, which usually causes public complaint.

Pentachlorophenol (PCP) was widely used through the textile industry as a biocide to prevent mildew and it was present in relatively high concentrations in this effluent. Since its inclusion on the list of controlled substances, all the firm's customers have been contacted with the aim of eliminating it from the supply chain; the figures demonstrate the success of this approach.

6 Summary

It can be seen that some of the parameters are very close to the current consent limits, and the plant continues to be modified and added to in order to comply with the legislation. Some of the salient points that contribute to the effective operation of the plant are summarised below.

(a) The site is composed of a large number of individual dyeing machines in which a cross-section of dyeing and finishing operations are performed. This factor and the installation of a balancing tank leads to a relatively constant stream of effluent to the biological treatment tanks.

(b) Unlike municipal treatment tanks, which operate at near-ambient temperatures, the biological stage is carried out at about 40 °C throughout the year. As a result, the BOD is removed within a few hours, with obvious economies in the size of the plant. The biomass must have developed a specialised thermophilic strain of bacteria to survive under these conditions.

(c) The incorporation of the sludge from the lime-soda treatment of river water into the effluent stream has several benefits. Apart from its ability to buffer out excesses of pH and precipitate any heavy metals, there is some evidence that it acts as an inert carrier for the biomass, increasing the surface area and helping to reduce the contact time. It also enhances the settling characteristics of the biomass, which in turn improves the efficiency of the clarifiers and keeps the plant relatively compact.

(d) Removal of the residual colour from the biologically treated effluent is relatively simple, using a cationic agent to precipitate the dye as a stoichiometric reaction. This additional solid, and any carryover of biomass from the previous stage, is effectively removed by dissolved air flotation.

There is no doubt that Stevensons made a wise decision in the 1960s to install an effective effluent treatment plant, and it has been possible to add processes using new technology to improve its performance. The plant is a valuable asset of the business and it has been estimated that it would cost more than £3 million at today's prices to build such an installation.

CHAPTER EIGHT

The treatment of dyehouse effluent at John Heathcoat

Eric J Newton

1 Introduction

Textile dyeing and finishing effluents are by and large highly polluting and generally extremely variable in content. The major pollutants in respect of chemical and biological oxygen demand (COD, BOD), and therefore those that have a major effect on the ecology of the watercourses into which they are discharged, are oils, waxes, detergents, sizes and carriers.

Dyes, whilst not producing any significant biological load on the watercourse, could be claimed to be nonpolluting. They are, however, highly visible in very small concentrations, and as such are far more likely to be the subject of complaint by an increasingly environmentally aware public than some other chemicals that can inflict real damage on the river ecology. Situated in south-west England, and having a dyehouse which makes considerable use of reactive dyes, John Heathcoat & Co. Ltd were among the first companies to come under pressure from the regulators to remove colour from their direct effluent discharge to the river.

The reasons for the demands were these:
- the company was discharging into the River Exe, a salmon and trout river, and one of the major salmon fisheries in the region
- the river is used as a drinking water conduit from the reservoir on Exmoor to supply the city of Exeter
- numerous leisure activities, such as coarse fishing and canoeing, centre on the river
- the river has very low flow characteristics during dry weather, due to the poor water retention of the soils in the area
- the river is situated in the middle of a picturesque holiday area.

2 Methods of colour removal

At the time the plant was to be installed (1988), knowledge of colour removal in bulk from dyehouse effluent was extremely limited. It was, however, known that conventional biological treatment plants had little or no effect on the colour of the effluent. From personal experience of running a large activated sludge treatment plant at a dyeworks in the south-east, it was known that the best that could be hoped for was the adsorption of a small amount of dye on the biomass. Some other method of colour removal had therefore to be found.

The methods considered were these:

– chlorination

– ozone treatment

– chemical precipitation (flocculation)

– special filtration techniques, such as reverse osmosis

– treatment with activated carbon.

Chlorination was found to give good colour removal, and there were methods available for the detection of free chlorine in the final discharge. The method was thought to be relatively inexpensive and had the advantage of having no waste products that would have to go to landfill. However, the NRA were known to be extremely sensitive to the possibility of even very small amounts of chlorine being present in the effluent discharge, since salmon and related fish are known to be severely affected by chlorine.

Additionally, the action of chlorine on complex dye molecules can produce chlorinated phenols, other chlorinated cyclic structures or split molecular fractions, all of which would be highly undesirable in a river that was eventually to be used for the supply of potable water. This possibility was therefore rejected.

Ozone treatment was at the time in its infancy, and it was believed that the high cost of producing the ozone would make the process highly expensive. Ozone treatment also suffered from the problem that molecular fractions of the dyes, some of which may possibly be toxic, would remain in the treated effluent.

Special filtration techniques using advanced membrane technology had the advantage of producing a final effluent that could probably be recycled, but the high volume of output required (up to 3000 m^3 per day) made the unit too expensive.

Activated carbon treatment was rejected because of the large amount of carbon that would be needed to decolorise a very highly coloured effluent, and because the cost of regeneration of the activated carbon was likely to be very high. This therefore left chemical precipitation as the only economically viable means of decolorising the effluent.

3 Chemical precipitation process

Extensive pilot-scale trials revealed some major advantages:
- the flocculation process was very effective in removing anionic detergents from the effluent stream, although it had (as expected) little effect on nonionic detergents
- excellent removal of the more troublesome reactive dyes was obtained; the concentration of flocculant was increased at times when the effluent was very highly coloured, and this dealt effectively with the extra colour
- the colour was removed completely, with little or no danger of forming toxic or foul-tasting residues.

Perceived disadvantages were that the flocculation is a little difficult to control and the rate of precipitation and size of floc could be affected by impurities such as nonionic detergents remaining in the effluent. Above-normal concentrations of reducing or oxidising agents were also found to affect the efficiency of the flocculation process, and these chemicals had to be neutralised in the dyehouse before discharging to the collection tanks.

The major disadvantage is that a sludge is produced which has to be settled, dewatered and pressed into a cake for subsequent landfill tipping. Some balancing tanks had already been installed, together with a settlement tank fitted with a rotary scraper bridge, and it was decided to design the plant round this existing installation. A three-stage chemical treatment unit was built, based on three large tanks (Figure 1). The mixed effluent is pumped from the balancing tanks to a coarse prefiltration unit to remove any fibrous matter. The effluent then enters the first tank, where it is dosed with a 10% lime slurry. This slurry is automatically mixed at the bottom of a large storage silo, and pumped to the first stage treatment tank. Dosing is controlled using an automatic valve connected to a control pH meter to give a pH of 11.3.

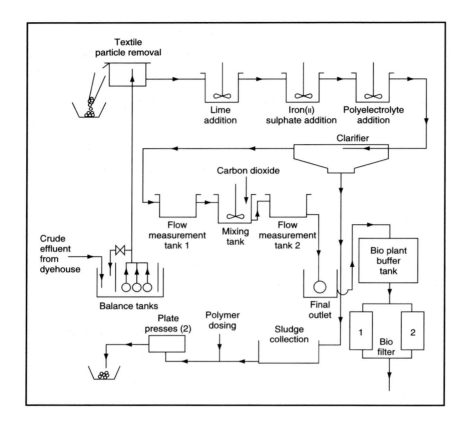

Figure 1 *Effluent flow diagram*

Each tank contains a large low-speed mixer which thoroughly mixes the effluent before it passes to the next stage. The mixers also prevent early settlement of the floc, but the slow movement of the paddles is insufficient to bring about breakdown of the floc into smaller particles.

Having reached a pH of 11.3 the effluent flows to the second tank, where it is dosed with an iron(II) sulphate solution, forming a precipitate of iron(II) hydroxide. The mixture then passes to the third tank where polyelectrolyte is added to aggregate the floc particles, enabling a faster and more effective settlement to take place in the settlement tank.

In the settlement tank the floc falls to the bottom of the vessel, where it is scraped to a central sump and automatically pumped to a sludge tank using a vacuum extraction device. The supernatant liquors weir over the top of the settlement tank and pass through two further tanks, where the pH is checked and the liquor brought to a pH of between 7.5 and 8.5 by injecting carbon dioxide under pressure, using a process developed in conjunction with ICI in which dosing is carefully controlled by a dual pH meter system. This process has superseded the original method involving pH adjustment using concentrated hydrochloric acid; this was difficult to control with sufficient accuracy and also gave problems with pumping the concentrated acid. A further advantage is that any accidental overdose with carbon dioxide would not drop the pH below 5–6, which is still satisfactory for discharge to the river, whereas an overdose of hydrochloric acid would almost certainly result in a fish kill and ecological damage to the watercourse. The carbon dioxide dosing system also has appreciably lower running costs than dosing with mineral acid.

The pH-corrected effluent is then pumped into a 300 m³ balancing tank for further mixing (to avoid peaks of BOD) before being pumped to the biological plants for further treatment. The sludge which has been removed from the bottom of the clarifier is allowed to settle further, and the supernatant liquors returned to the raw effluent balancing tanks to be retreated through the chemical treatment process (not shown in Figure 1).

The thickened sludge is then further conditioned by treatment with a different polyelectrolyte, which makes it more granular; it is thus easier to express the water by filter presses, forming a cake which is then sent to landfill. The expressed liquors, whilst clear, have been found to be very high in BOD and are therefore returned to the raw effluent tanks to be re-treated through the chemical process.

The process is successful in decolorising strongly coloured dyehouse effluent (Figure 2) and there are very few dyes that cannot be handled effectively. Effluent treatment, however, cannot be regarded as a separate and distinct process from the dyeing unit itself. The process *must* be viewed as a whole, including the effluent treatment section, since there are many ways in which viewing the process as a whole can be of help to the dyehouse manager.

For instance, nonionic detergents, if present in a reasonably high concentration, can interfere with both floc size and settlement rate, and it is therefore advisable where possible to keep detergents of this type to as low a level as possible commensurate with safe processing; alternatively, they may be replaced with anionic detergents capable of doing the same job.

Hydrogen peroxide can oxidise the iron(II) content of the floc to iron(III), which, whilst still giving reasonable decolorising, gives considerable problems with floc aggregation and settlement, and it is thus advisable to 'neutralise' the peroxide before it is released from the dye vessel. Large quantities of reducing agents can also interfere with the precipitation and

Figure 2 *Raw effluent (left) and final treated effluent*

settlement process, and are best at least partially dealt with before being sent down the effluent drain.

The whole unit is monitored for pH throughout and the process is automatically stopped if the pH goes outside a previously determined limit.

4 Treatment of remaining soluble BOD

The chemical precipitation process removes in excess of 50% of the BOD of the raw effluent, but the decolorised effluent has still too high a BOD loading for discharge into the river.

Some form of further treatment is therefore required, and in this case it is better that the treatment is of a biological nature. There are several considerations to be taken into account when choosing a suitable plant for biological treatment, and these are discussed in turn below.

4.1 Space

Conventional biological treatment plants consisting of an aeration basin and a settling and sludge return tank, whilst undoubtedly effective, require a large amount of space. The Heathcoat works were critically short of space in which to install a large plant (the same is probably true for most other dyehouses), and a properly sized conventional activated sludge unit could not be fitted into the space available.

4.2 Variability of the effluent

Effluent from a commission dyehouse is notoriously variable in composition, according to the types of fibre being dyed at any one time, and our testing showed massive variations in BOD load from hour to hour; this is something which biological units are not very good at dealing with.

4.3 Flow variations

Biological plants work best when they are used under constant flow conditions.

4.4 pH

For optimum efficiency and to prevent kills of the biomass, it is best to keep the influent to the plant within the bounds of 1.0 pH unit. Having looked at the variability of the BOD loading of the effluent coming from the chemical treatment plant, it was decided that some form of intermediate balancing capacity was a necessity, and a tank of sufficient size was installed to be capable of holding a volume equivalent to 4 hours average effluent from the dyehouse. This has the following advantages.

(a) It provides further blending of the chemically treated effluent, and thus a more consistent biological load to the bio plant.

(b) There is a reservoir of up to 4 hours of effluent, to cover the times of the day when the dyehouse is discharging very little to drain.

(c) The chemically treated effluent can contain small amounts of fine iron(II) hydroxide floc which has not settled properly. In the event that a submerged filter biological unit was chosen, the presence of the holding tank would give this precipitate further time to settle, and thus avoid early blockage of the filters of the biological unit.

5 Choice of plant

Having defined the parameters which the proposed biological plant had to meet, the market was then comprehensively surveyed for a suitable unit. These investigations led to the choice of a plant built and marketed by PWT Projects Ltd of Hounslow, Middlesex. The unit was based on technology developed by PWT's parent company Thames Water plc, and had been designed specifically to treat domestic effluent for small to medium-sized communities where sewerage connection to a large central plant was difficult. It had never been used on an industrial effluent, but there did not seem to be any major reason why it should not work effectively, providing the effluent to be treated was reasonably biologically soft.

The plant is based on the use of submerged aeration filters, and has the major advantage that it occupies only around 20% of the space of a conventional activated sludge plant of a similar capacity. The unit has no need of a settlement tank, and the biomass grows on beds of expanded porous shale arranged in compartments overlying special aeration units, which bubble air at a high rate through the shale.

The chemically treated effluent is introduced into the top of each compartment and percolates down through the shale beds against the flow of air, with the biologically treated and filtered effluent coming out of the bottom of the filters into a built-in holding tank, before discharge to the river (Figure 3).

In order to prevent excessive build-up of biomass on the shale, the beds are backwashed at specific intervals using biologically treated water stored in the unit's holding tank. The biosludge removed by the backwashing is returned to the chemical plant to go round the flocculating process again and eventually ends up being mixed with iron(II) hydroxide sludge, to be dewatered and filter pressed for landfill.

Air for the process is provided by quiet high-volume compressors in specially insulated housings, which present no noise problems even running at night in an area close to domestic housing. The plant is totally automatic and computer-controlled, requires very little operator attention and runs with an almost complete absence of unpleasant odours: again, this is important where domestic housing is nearby.

The plant can be put on stand-by mode to enable the biomass to withstand a prolonged holiday shutdown. Normally sufficient effluent is kept back in the balancing tanks to run the plant for 15–20 min every 24 h during the shutdown, and the plant will usually return to its

Figure 3a *(Left) Two of the biofilter compartments showing one compartment backwashing and the other in normal operation; (right) overall view of one of the biofilters*

maximum working efficiency within 24–36 h of the re-start. If it is not possible to preserve any effluent, the plant can be dosed with water containing concentrated nutrients, with the best mix being determined on a trial-and-error basis.

Because of original uncertainties caused by the fact that a plant of this type had not operated previously on an industrial effluent, one unit was installed to cope with 50% of the works' requirement, followed by a second 15 months later when it was clear that the first unit was capable of consistently delivering effluent of the desired quality.

6 Costs

The total cost of the installation, both chemical and biological, was £1 500 000. The biological units represent a one-off capital investment cost, and the running costs are low.

The chemical treatment plant, however, incurs quite high running costs both in chemicals and for landfill. There is thus a need for an even-handed approach by the regulators across the UK, and probably even more importantly with our European and foreign competitors, to prevent as far as possible localised cost distortions caused by differing regulatory requirements. It is little use having the finest environmental process if the goods produced are unsaleable because of competition from suppliers who have not had to meet the same standards.

7 Where to now?

An immense amount of research has been carried out into the treatment of coloured textile effluents, and the direction for this research is reasonably clear.

Costs of landfill are certain to escalate as tipping regulations become more stringent and the availability of suitable holes in the ground (not too near anyone's backyard!) diminishes. Treatment methods for the removal of colour must therefore concentrate on reducing or eliminating the production of sludge, thus reducing the cost of landfill disposal. If a biological method for removing colour on a large scale can be found, then this must be the ideal solution. Use of other flocculants, in conjunction with special polyelectrolytes, offer some promise, with the precaution that many polyelectrolytes if used in excess can have an inhibiting effect on any subsequent biological treatment.

Special filtration methods may be developed to produce re-usable water, but there remains the problem of what to do with the concentrated impurities.

The dye manufacturers may be able to help by producing dyes (particularly reactive dyes) with a better fixation than can be achieved at present, although developments in this field are believed to be still some way off. The need is urgent and immediate, and pressure must be maintained to evolve cheap and effective methods of colour removal.

8 Conclusion

The industry has little alternative but to respond to the environmental pressures from a public which is being clearly made more aware of the need to protect the environment in which they are living. The pressures will not go away, and standards will be enforced with increasing severity by the regulators.

Work to provide more cost-effective methods of treatment of colour will continue and it is to be hoped that in the short term a more environmentally aware public will be prepared to pay a little more for fabric that has been produced in a manner in which care for the environment has been a major factor.

Part Three Currently available technologies

Chapter 9

Macrosorb colour treatment systems, *Keith R F Cockett and Maurice Webb*

The effluent treatment system described is built around the absorbent properties of the particulate inorganic material Macrosorb. This absorbs dyes and organotoxins from process waters down to levels close to detection limits, and is then removed by accelerated gravity settlement. Treated effluent is substantially colour-free, and has a reduced COD and a much reduced mothproofer and pesticide content. This permits its discharge to river or sewer within current and foreseeable consent limits and, in some instances, its re-use in the dyehouse, which can lead to recovery of capital and operating costs.

Chapter 10

Arcasorb D dye effluent treatment process, *Philip C Blowes, Anne Jacques and David Jones*

A fixed-bed adsorbent process has been developed by Archaeus Technology Group to remove soluble dyes (including reactive dyes) and other contaminants from dyehouse effluent in a compact and economically viable operation. Trials at dyehouses in Britain and Europe have demonstrated the operational capacity of the process and detailed cost analyses have been prepared. Contracts for full-scale plant are currently being negotiated.

Chapter 11

Treatment of dyehouse effluent for subsequent re-use of reclaimed water, *Peter H Weston*

With rising costs of municipal water treatment, sewerage and waste disposal, together with escalating environmental concerns, competent waste management is essential. For small waste water generators, however, the expense of compliance with consent conditions is quite substantial, and treatment alternatives are somewhat limited. It would seem, therefore, that a feasible alternative might be in-house treatment. A point may exist at which in-house (end-of-pipe) treatment and re-use of the water reclaimed begin to outweigh conventional usage and discharge costs.

Chapter 12

Membrane technology for the separation of dyehouse effluent, *Clifford Crossley*

The article describes the general principles and theory of crossflow membrane filtration, together with the differentiating factors of the four processes, flow diagrams, membrane configurations and typical systems configurations. Membranes for colour removal from dyehouse effluents and associated liquors have been evaluated. The trials data on different treatment processes are discussed, as are the factors that influence a full-scale plant design. These factors include process selection, size (area), membrane type or configuration, operating costs and cleaning frequency. Recent increases in consent limits for coloured effluent disposal and water costs in the UK have forced industry to evaluate the available effluent treatment technology including membranes. Membranes will not solve the colour problems, but they offer a means to a financial payback by water re-use and effluent volume reduction.

Chapter 13

Colour removal from dyehouse effluents using synthetic organic coagulants, *Peter J Hoyle*

Appropriate dyehouse effluent management enables the removal of dye residues using synthetic organic coagulants. Recent advances in polymer technology and application techniques have allowed removal of residual colour to the extent that final discharge of treated effluents meet quoted consent levels.

Chapter 14

Effluent treatment using chemical flocculation, *Brian W Iddles*

Much research has been carried out on novel treatment chemicals. Several simple but effective systems have been developed, and a complete package can be offered to clients. The chemicals used are simple inorganic compounds with a small amount of polyelectrolyte. Clients are informed as to the nature of these chemicals, and are then free to negotiate terms with a supplier of their own choice. The chemical package is specific to both that particular effluent and that particular site. It has great flexibility, and has been successfully used in both the UK and foreign markets in industries ranging from animal rendering to precious metal recovery.

CHAPTER NINE

Macrosorb colour treatment systems

Keith R F Cockett and Maurice Webb

1 Introduction

Macrosorb is the trade name for a range of inorganic particulate absorbents developed jointly by Crosfield and Unilever Research and manufactured by Crosfield in Warrington. It is classified as a 'synthetic inorganic clay' and the crystal structure consists of parallel layers of clay platelets which, due to their chemical composition, carry a net positive charge. Between these layered platelets are anions which balance the cationic nature of the clay.

Clays belonging to this family do occur naturally, although they are not common and are usually found in an impure state. Synthetic forms have found applications elsewhere, ranging from pharmaceuticals to catalysts. Macrosorb is a specific particulate form whose crystal structure and particle morphology have been engineered for optimum absorption of the contaminants found in dyehouse effluents: large, negatively charged or polar molecules. Most common dyes and pesticides such as mothproofers fall into this category. Macrosorb can remove such molecules from aqueous streams rapidly and it has a high capacity. Its most valuable feature is that it can remove them down to extremely low levels, which challenge both visual and instrumental detection limits. This offers the prospect of being able not only to meet any current or envisaged consent limits, but also of water re-use (one benefit of which can be that treatment is self-financing).

Absorbing a contaminant from solution on to a particle can only solve the effluent problem, however, if the absorbent can then be separated from the waste stream and disposed of safely, and if the overall treatment system is operable in a cost-effective manner. The system

113

which has been developed around this absorbent employs other techniques and equipment that are already proven in water treatment. This chapter will describe the different technologies involved and how they are brought together in the treatment system; the results quoted are based on actual dyehouse experience.

2 Techniques and technology

2.1 Absorbing the contaminant

The primary mechanism of colour removal is by means of an effectively irreversible absorption of the dye molecules by the Macrosorb particles. The absorption is thought to take place via a combination of physical adsorption of dye on to clay surfaces within the microporous structure of the particles, enhanced by an ion-exchange process wherein the interlayer anions of the clay are displaced by the dye molecules. Removal of contaminants can take place at any pH between 4 and 11 (below pH 4 the Macrosorb dissolves, and above pH 11 absorption becomes far less effective) and at any temperature between 0 and 100 °C. In the proposed process, Macrosorb is used at pH 5.5–6.5 and at the effluent temperature as received (generally 35–40 °C).

2.2 Forming a separable particle

Absorption is followed by flocculation, in which the Macrosorb particles are aggregated into a readily separable floc. Two processes are involved here: *in situ* precipitation of metal hydroxide, and 'caustic shock'. Both have been used quite extensively in water treatment to remove finely suspended and colloidal matter. In this system magnesium chloride solution is added to the waste stream, followed by caustic soda. As the pH is raised two things happen. Firstly, magnesium hydroxide separates from solution; the precipitate forms around the Macrosorb particles and effectively aggregates them. Secondly, the rapid pH increase (caustic shock) denatures organics in solution, which themselves form a floc, and also aids settlement of existing flocs. The flocculation process not only produces a dense, readily separable aggregate particle but also contributes towards removal both of colour and of other suspended organics, thus significantly reducing the COD. In some instances, flocculation can be aided by addition of very low concentrations of organic coagulants.

Optimisation of this synergy between the component chemical processes to best suit the particular waste stream minimises the overall raw material costs.

2.3 Solid–liquid separation

Several options have been considered for separation of the floc plus contaminant from the cleansed water, including:

– settlement

– flotation

– membranes

– filters

– inclined plane (plate) separators.

The criteria used in the selection of a suitable separation system were that:

(a) the plant must be simple, well proven, easily maintained and have as few moving parts as possible

(b) the equipment must occupy as small an area as possible and require the minimum amount of civil engineering work on site

(c) the system must be capable of being scaled up to cope with any required throughput

(d) the system must be cost-effective.

The following paragraphs discuss the available options against these criteria.

Gravity settlement

This floc system will now separate under gravity from the aqueous phase to leave an effluent which will be substantially colour-free, with reduced COD and a much reduced mothproofer and pesticide content. As such, the treated effluent, together with any residual heat left in the water, can be re-used in the dyehouse. Gravity settlement meets criteria (a) and (c) and, in some instances, may be a preferred option for solid–liquid separation. In general, however, a significant settlement time, and hence a large settlement tank, is required; the system will not meet criterion (b) unless such a settlement system exists already or unless space is not at a premium.

Flotation

The dense, settling floc produced in this effluent treatment system is not amenable to conventional flotation processes. One possibility might be dissolved air flotation, in which dissolved, compressed air is injected into a treatment tank and carries the floc to the surface, where scrapers then remove the solid sludge. This, however, fails to meet criterion (a) adequately (compared with the system eventually selected) because of the mechanical complexity of the system. The negative aspects are the greater number of moving parts and the comparatively high capital and running costs, particularly of the air compressor. A positive feature is that sludges are, in general, more compressed (higher solids content) than those deriving from gravity settlement.

Membrane systems

Membrane systems can be used successfully to effect floc–water separation; in fact, a Macrosorb-based system utilising a crossflow membrane is used to improve dyebath wash-off efficiency and efficacy. Although such a system meets criterion (b), however, it has several disadvantages when it has to cope with the much greater volume throughputs and lower dye concentrations encountered in treating effluents. Membrane systems are complex and labour-intensive, both in operation and in maintenance, and capital costs would also be higher for these high volumes. A further risk is that the high COD levels encountered in effluents could cause membrane blinding and thereby disrupt continuous operation.

Filters

During the development of this system, several variations on the dead-end filter have been examined. All of them suffered one conclusive disadvantage: the permeability of a bed of these fine particulates is so low that, in order to cope with the flow rates likely to be encountered, the filter surface area would need to be huge. Criteria (b), (c) and (d) could not therefore be met.

Crossflow clarifier

This is also known as an inclined plane separator or lamellar plate separator. It was chosen as the preferred solid–liquid separation device basically because it met criteria (a), (b) and (c) at the lowest capital and operating costs. The system (described in more detail in the next section) has no moving parts and is well proven. It is widely used for the separation of natural clays and silts from mineral-processing effluents, as well as for the removal of precipitated solids from industrial waste streams. Also it is readily available in a range of sizes to meet any volume throughput requirement. The separated sludge is easily transported from the base of the vessel and scrapers are not required. In addition, the compressive action during settlement produces a denser sludge than does simple gravity settlement. The crossflow clarifiers used in the Macrosorb treatment system have been modified by the manufacturers (Bergmann) to suit the settlement characteristics of the Macrosorb floc and are tailored to the individual dyehouse requirements.

The sludge, as described above, consists of contaminated Macrosorb, precipitated magnesium hydroxide and denatured organics produced by the caustic shock. A typical solids concentration at the base of the crossflow clarifier is about 1% dry weight solids. This sludge is pumpable, and one disposal option is to deliver it to the sewage treatment works. It can also be concentrated in order to minimise disposal costs – to landfill, for instance. Concentration is brought about by mechanical dewatering using, for example, a centrifuge, a continuous-belt filter or a plate-and-frame filter press. The last-named, the preferred option, gives a high-solids-content filter cake – typically 20–25% dry solids. At this concentration the filter cake is a hard, stackable solid.

3 Operational flow diagram

Figure 1 is a flow diagram of a typical Macrosorb treatment system. The equipment includes four tanks:

–   tank 1: reaction tank
–   tanks 2 and 3: flocculation and coagulation
–   tank 4: crossflow clarifier.

4 Operating principles and system description

4.1 Stage one

The balanced effluent is automatically dosed, in line, to the correct pH for effective

117

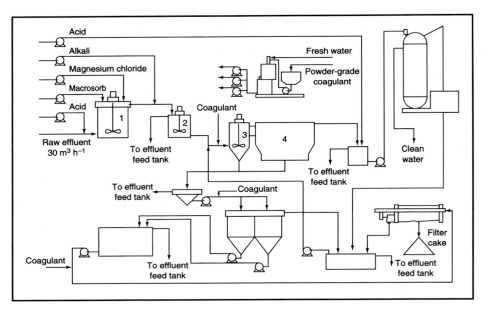

Figure 1 *Macrosorb treatment system flow diagram*

Macrosorb treatment. Analysis of composite samples supplied by various dyehouses suggests that the need for such acid dosing is minimal, and it may on occasion be unnecessary. Concentrated acid is used as this is less expensive than dilute acid but, in order to avoid operative handling on site, the acid is automatically diluted immediately prior to addition to the line. Typically, the pH in tank 1 is 5.5–6.0.

Macrosorb slurry and magnesium chloride solution are dosed to tank 1, which is stirred in order to give efficient contact. The tank size is adjusted to give sufficient residence time to ensure effective absorption of the contaminants on to the Macrosorb, covering the range of flow rates likely to be encountered. The treated liquor then passes to tank 2 using hydrostatic pressure (head) as the motive force.

4.2 Stage two

During the passage from tank 1 to tank 2, caustic soda is dosed into the flow causing a rapid increase in pH, typically from about 5.5–6.0 to 11.0–11.5. As a result, magnesium hydroxide precipitates, and a floc is produced from the dissolved organics (section 2.2). Both reactions increase the size of the suspended particles and hence the rate at which these particles settle.

A short residence time in tank 2 allows for completion of these reactions prior to passage on to tank 3. Again, the reagent (caustic soda) is used in concentrated form for economic reasons and, again, the concentrated reagent is diluted by a water flush immediately prior to addition.

4.3 Stage three

In tank 3 a suitable, high-pH-stable, high-r.m.m. coagulant is added via an aqueous flush. Flow into tank 3 is tangential, giving adequate coagulant mixing whilst preserving the coherent nature of the floc. At this stage the floc is sufficiently large to allow for a degree of sludge separation from the bottom of tank 3. This sludge is combined with that from the crossflow clarifier.

4.4 Separation

The crossflow clarifier, which removes the bulk of the contaminant-loaded Macrosorb and other solids from the effluent, is widely used elsewhere as a high-rate gravity separation device. An array of closely spaced inclined plates within a proprietary cell configuration provides a large settling area for solids in suspension. Particles settle a short distance only before contacting one of the plates, at which point they are removed from the effluent flow which, now relatively solids-free, is discharged. Accumulated particles slide down the plates, falling into quiescent channels in the cell and from there to the base of the tank to form a sludge. There are, therefore, two outputs from the crossflow clarifier: an essentially solids-free effluent and sludge.

4.5 Final filtration

In cases where the final effluent consent standards for residual contaminants are low, or where the treated effluent is to be re-used in the dyeing processes, an additional stage of treatment may be required.

The crossflow clarifier should consistently reduce the level of suspended solids present in the pretreated effluent stream to within the range 50–100 mg l^{-1}. In some wool dyeing sites, where occasional high permethrin levels are combined with very stringent effluent discharge standards, the suspended solids still present in the treated effluent may contain permethrin in excess of consent levels. Tests have established that an additional separation step to remove solid particles down to 10 mm size will reduce the residual permethrin concentration to acceptable levels. A continuous sand filter is used for this purpose.

The effluent from the crossflow clarifier collects in a balancing tank, whence it is pumped into a vessel containing a deep bed of graded sand particles. As the liquid filters down through the pores in the sand bed, fine solids are retained, allowing a final treated effluent to be discharged to sewer (or considered for re-use in the dyeing processes). As retained solids build up within the sand bed, the pressure drop across the filter increases until a point is reached where backwashing is required. An automatic sequence is initiated, wherein treated effluent is forced into the sand bed in the opposite direction to the normal flow. This flow fluidises the bed, allowing the retained solids to be released. Backwash water and solids are returned to the treatment system and the filtration process is resumed.

4.6 Aftertreatment pH correction

The Macrosorb effluent treatment system uses raised pH levels to encourage floc formation. Automatic post-treatment pH correction is included to bring the pH of the final treated effluent within consent limits.

4.7 Sludge treatment

Having separated the contaminant-loaded Macrosorb and other solids from the effluent stream, the resulting sludge requires treatment before disposal. Accumulated sludge in tank 3 and the hopper bottom of the crossflow clarifier is periodically discharged under timer control to a collection sump. Sludge is pumped from the sump to a sludge thickener comprising one or more cylindrical, conical-bottomed tanks with slowly rotating rake and agitator assemblies. With the aid of further flocculant or coagulant addition, the sludge consolidates, allowing a clear supernatant to overflow from the top of the sludge thickener for return to the treatment process and a thickened sludge to discharge from the bottom periodically under timer control. The volume of sludge from the treatment process can thus be reduced by a factor of between two and three.

Thickened liquid sludges can be disposed of, but the volumes involved and the liquid nature of such sludges give rise to relatively high transport and disposal costs and a restricted number of receiving sites. As a consequence an additional sludge treatment step, mechanical dewatering, is included to achieve a further volume reduction and a more acceptable product for disposal. Several mechanical dewatering processes have been evaluated, and a filter press has been selected that is able to produce a readily handled product with a volume-reduction factor of between 7 and 12.

The thickened sludge is pumped from the sludge thickener to a large balancing tank, which is required to allow batch operation of the filter press during normal working hours. At the start of the dewatering cycle sludge, which may be conditioned by further flocculant or coagulant, is pumped to the filter press.

The filter press comprises an array of recessed plates within a frame, into which the conditioned sludge is fed. Filter cloths are held between each pair of plates. As the pressure increases the chamber fills, water passes through the cloths and discharges via ports in the surrounding plates, while the retained solids progressively dewater. At the end of the dewatering cycle, an operator opens the array of plates, allowing the dewatered filter cake, with an anticipated dry solids content of about 20%, to discharge into a skip or other receptacle. The filtrate is returned to the treatment process.

4.8 Water re-use

The treated effluent produced from the above Macrosorb system has a reduced COD, is substantially free of colour and could be re-used in several dyehouse processes. Because of the relatively short treatment times required, the treated effluent still has a worthwhile heat content; should it be re-used, then the heat content in that portion of the water recycled is also available.

4.9 Sludge disposal

At present, one of the options available is to pump the undewatered sludge from the bottom of the crossflow clarifier to the drain for delivery to the sewage treatment works. This may be the most cost-effective option, particularly if there are no toxic contaminants removed by the Macrosorb treatment system. Alternatively, the sludge can be disposed of as liquid effluent without further treatment, or it can be dewatered further as described above, preferably in a plate-and-frame filter press, then disposed to landfill.

At present, it is not economically feasible to strip absorbed species from the absorbent to permit absorbent re-use. Water-based stripping options would, in any case, regenerate a solution of colour and other contaminants, albeit a more concentrated one than the original effluent. Research is under way, however, to evaluate options for re-use and alternative use of the spent absorbent which, if successful, could increase the environmental attractiveness of the Macrosorb treatment system still further.

4.10 Balancing

Irrespective of which effluent treatment system is adopted, the treatment must be preceded by an adequate effluent balancing system. Without this, treatment dosages would need to be wastefully high in order to ensure adequate treatment of the occasional peaks occurring during production. The volume of the existing effluent tanks on many sites is adequate, within that which is normally recommended for the Macrosorb effluent treatment system. Often it will be necessary only to carry out some relatively simple and low-cost modification to the tanks currently in use. It is, however, recommended that a mixing specialist is consulted to determine final design arrangements.

5 Results obtained using the Macrosorb treatment system

The following information has been compiled from a considerable body of laboratory and field trials, using both real and simulated effluents.

5.1 Dye types

The human eye can detect concentrations of some dyes down to extremely low levels (as little as 0.01 ppm in a clear river), particularly in the red and purple regions of the spectrum. Currently even a well-run dyehouse can be expected to discharge effluent water which is significantly coloured. Information on which dyes can be treated with the Macrosorb system is as follows.

(a) *Acid dyes* These can generally be readily removed. Certain low-r.m.m., highly sulphonated dyes require special techniques.

(b) *Metal-complex dyes* 1:2 and 1:1 metal-complex dyes with varying degrees of sulphonation or carboxylation are readily absorbed.

(c) *Chrome dyes* In the unchromed state, chrome dyes are absorbed by Macrosorb. In the chromed state such colouring matters are not in solution but are removed by the accelerated gravity settlement of the treatment system.

(d) *Direct dyes* These are readily removed.

(e) *Reactive dyes* These dyes are generally regarded as the most difficult to remove from textile dyeing effluents, because of their chemical nature and high concentration. Macrosorb effectively removes such dyes irrespective of the reactive group (vinyl sulphone, chlorotriazinyl etc.) and irrespective of whether the dye molecules contain one or two reactive groups.

(f) *Disperse dyes* These are only slightly soluble in water but a combination of absorption and gravity settlement in the overall system gives good removal.

(g) *Azoic, vat and sulphur dyes* When present in effluent, such dyes are actually pigments rather than solutions, but they are found in consolidated dyehouse effluents in mixtures with other dye classes. The Macrosorb system copes well with such mixtures.

5.2 Other contaminants

(a) *Adsorbable organohalides (AOX)* These are produced by chlorination of cellulosic materials and by the use of chlorine pretreatment in wool shrinkproofing. The system removes both soluble and insoluble AOX entities.

(b) *Pesticides and mothproofers* Pesticides generally derive from natural fibres and are fairly low-r.m.m. organics with only slight aqueous solubility. Macrosorb can remove a wide range of organochlorine and organophosphorus pesticides, including the frequently encountered sheep dip chemicals propetamphos and diazinon. Narrow-spectrum pesticides, such as the mothproofers permethrin and cyfluthrin, are also removed. Even with the best possible housekeeping practice, it is difficult to keep mothproofer levels in the effluent from current plants to much below 50–100 ppb. Severn Trent and Yorkshire Water Authorities, which cover the areas of the UK where most dyers are located, are in the process of imposing reduced consent levels.

(c) *COD* As well as removing contaminants such as the colours and toxins listed, Macrosorb can also effect significant reductions in COD.

5.3 Problem contaminants

So far, Remazol Yellow R is the only dye found to pose a significant problem to the standard sorbent system (although there may be others of a similar chemical nature, i.e. small, highly hydrophilic molecules). The removal of this colour can be improved by a modification to the sorbent system.

5.4 Typical results

Tables 1 to 4 show results derived from treating real effluents and illustrate the levels of improvement which can be obtained with this system. Figure 2 shows results from a field trial taken over a continuous operating period of 48 h, in terms of average absorbance values over the range 400–700 nm. Experience on wholly cotton-dyeing effluents suggests that an average COD reduction of up to 75% can be achieved, which constitutes another worthwhile cost-saving element in downstream treatment.

Table 1 Improvement in colour achievable with Macrosorb system

	Absorbance at 100 nm intervals			
	400	500	600	700
Before treatment	0.59	0.89	0.23	0.11
After treatment	0.025	0.008	0.005	0.000
NRA objectives	0.025	0.015	0.008	0.003

Table 2 Improvement in mothproofer content achievable with Macrosorb system

	Mothproofer content/ppb	
	Permethrin	Cyfluthrin
Before treatment	31	4
After treatment	<0.27	<0.1
Consent limits	2.5 to 8	

Table 3 Improvement in pesticide content achievable with Macrosorb system

	Pesticide content/ppb	
	Propetamphos	Diazinon
Before treatment	26	1.5
After treatment	0.9	0.1

Table 4 Improvement in COD achievable with Macrosorb system

	COD/mg l^{-1} a
Day 1	
Before treatment	1635
After treatment	715
Day 2	
Before treatment	1920
After treatment	676

a Mixed cotton- and wool-dyeing effluent

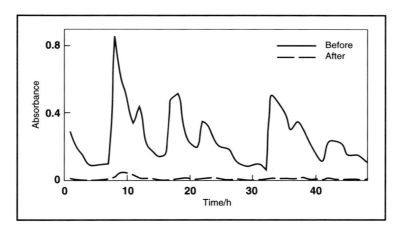

Figure 2 *Comparison of untreated and treated effluent over a 48 h period*

5.5 Water re-use

Trials have been carried out using 100% treated effluent to dye and finish fabrics. The finished article was indistinguishable from those processed using fresh water. Depending on the level of water re-use, a build-up of contaminants is possible, however; it is therefore suggested that a level of 50% water re-use should be targeted.

6 Costs and savings

Treatment costs are minimised by tailoring the equipment and the chemical suite to best suit the requirement of an individual process or plant. Requirements and economics of operation will clearly vary from site to site, but experience to date indicates that not only is this treatment system cost-competitive, but that in many instances where effluent water from dyehouses can be re-used, cost savings are possible. A five-year leasing facility is available, offering advantages in taxation and allowing the equipment to be financed from income rather than from capital.

7 Environmental benefits: now and in the future

Environmental and economic benefits do not always accompany each other but, in many respects, this is the case for the Macrosorb treatment system. The most pressing requirement is for colour and toxin discharge standards to be met by effluent water streams, and this system can technically meet any standards likely to be encountered in the foreseeable future. The system also permits at least partial re-use of water, together with any associated heat and chemicals, thus reducing the volume of waste delivered by the dyehouse and reducing its water off-take. In addition, it reduces COD in the waste stream.

Macrosorb itself is environmentally harmless and there are no hazards associated with its manufacture and use, nor even with its 'misuse'. (For example, overdosing of some treatment chemicals can cause problems downstream either for the sewage treatment plant or for the consumer and, as far as the dyehouse is concerned, could prohibit water re-use even if all discoloration were removed.) As far as disposal is concerned, the only hazards are those inherent in the absorbed contaminants.

For the future, there are three lines of investigation which could lead to further environmental benefits, as follows.

(a) Research is already in hand to increase the weight-effectiveness of the Macrosorb. The benefits from success here will be reduced sludge volumes and smaller treatment units, consuming less energy and chemicals.

(b) It would be useful to find environmentally better ways of disposing of, using or even recycling, the absorbent. This could reduce raw material demands and energy associated with manufacture.

(c) It may be possible to design absorbent systems that remove colour more selectively, thus permitting water recycling, but which do not absorb other benefit chemicals in solution, which might thereby be returned to the process; it seems unlikely, however, that the colours themselves can be recycled.

Thus this treatment system already goes a long way towards cleaner technology and holds out promise for further steps in that direction.

CHAPTER TEN

Arcasorb D dye effluent treatment process

Philip C Blowes, Anne Jacques and David Jones

1 Technology type; other uses in similar situations

Arcasorb D is a proprietary biological adsorbent (biosorbent) from Archaeus Technology Group[1] which removes soluble dyes (in particular reactive dyes) and other contaminants from textile waste streams. The biosorbent is derived from a naturally occurring polymer. The raw material from which it is manufactured is already available in tonnage quantities as a waste product from the food industry. Arcasorb D is supplied as a granular solid and is operated in a packed bed system which allows a compact plant design. Effluent treated with Arcasorb D may be discharged directly to sewer (with pH correction as necessary) or passed forward to a reverse osmosis membrane system to produce full-recycle-quality water.

The Arcasorb D bed is regenerated *in situ* using dilute chemicals at intervals determined by the volume and flow rate of effluent to be treated, the average daily colour of the raw effluent and the level of colour removal required to achieve targets for either discharge to sewer or recycle. The resulting concentrated regeneration liquor is flocculated and the solids disposed of to landfill as a semisolid filter cake. The resulting clarified liquor may be returned to the front of the process to repass through the Arcasorb. If the liquor is sufficiently clarified, however, it may be passed forward to a reverse osmosis system and go on to be recycled or else discharged direct to sewer, depending on consent limits or client requirements. This compact system means that the plant footprint is small and can easily be accommodated in the limited space available in most dyehouses.

---

[1] Arcasorb effluent treatment systems are now built and marketed by Arcasorb Technology Ltd (telephone 01923 711587)

A related product, Acrasorb 200, will remove coloured compounds such as humic and fulvic acids from upland water sources, together with associated metals such as iron, aluminium and manganese, and has been approved for use in the treatment of drinking water for public supply. Arcasorb products can also remove total organic carbon and variable amounts of chemical oxygen demand (COD), in addition to certain pesticides, such as the mothproofing agent permethrin, from liquid streams. In laboratory trials Arcasorb products have also removed phenols, cresols and other organics from industrial effluents.

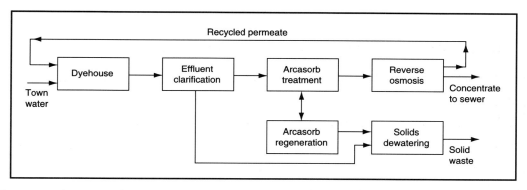

Figure 1 *Dyehouse water recycling scheme using Arcasorb D*

2 Operational flow diagram

The block flow diagram of the Archaeus dyehouse water recycle scheme is shown in Figure 1. Three principal operations are performed on the dyehouse effluent:

– clarification

– Arcasorb treatment

– reverse osmosis (RO).

Also provided are Arcasorb regeneration and solids dewatering systems.

2.1 Clarification

Entire effluent is lifted by a pump from the dyehouse and passed into the clarification system via one or two hours balancing capacity. Equipment is then provided to remove in turn:

- larger adventitious items, such as parts of garments
- small particles, such as fibres shed from textiles
- very fine particles and colloids, such as chemical precipitates and spent dyes
- emulsified organics such as lubricants.

The effluent is first passed through a coarse screen, to prevent blockage of the downstream finer-filtration equipment, and is then subjected to pressure filtration for optimum clarification at an economic cost.

The solids separated from the effluent pass to the dewatering equipment and are combined there with the regeneration slurry (see below) to produce a solid waste for disposal (to landfill, for example). Acidification of the effluent, required for the Arcasorb treatment, is also carried out in the clarification section.

2.2 Arcasorb D treatment

Arcasorb D has been developed so that it can be employed in equipment styles familiar in water treatment processes, such as sand filters, softening or ion exchange.

A typical Arcasorb D contactor contains a bed of Arcasorb 1–2 m deep. Service flow of effluent is downwards. At the approach velocities used pressure loss is low, since the bed voidage is high; this high voidage avoids the risk of plugging by any suspended solids slipping past the upstream filtration. The colour breakthrough point is typically set at 5 or 10% of the design level in the raw effluent.

A dyehouse working 24 h per day will require more than one bed of sorbent, so that treatment of effluent may continue while a bed saturated with colour is regenerated. Typically two beds, connected in parallel, are provided; larger-capacity dyehouses may require more than two.

Arcasorb D regeneration

An exhausted bed is isolated from the treated effluent header. The adsorbed colour is stripped off the Arcasorb by passing a flow of high-pH regeneration liquor through the bed in the counter-current direction. The initial regeneration liquor can be made up using raw dyehouse effluent, with the addition of further chemicals.

The used regeneration liquor, containing highly concentrated colour, is neutralised with acid and flocculant is added. Solid colour material separates and, after settling for an hour, the slurry is sent to the solids dewatering equipment. The supernatant liquor colour is usually low enough for it to be passed forward to join the treated effluent. After sufficient colour has been stripped from the Arcasorb D bed the regenerant pH is reduced by the addition of acid so that the medium is restored to the optimum pH for colour adsorption. The freshly stripped and re-acidified bed is reconnected to the treated effluent header to start another adsorption phase.

Solids dewatering

The solids removed by the effluent clarification equipment, together with the colour materials precipitated from the regeneration liquor, are transferred to equipment where the solids:water ratio is increased. The resulting cake is transferred to a skip for periodic removal by a waste disposal contractor to landfill.

2.3 Reverse osmosis

When the effluent from the dewatering unit is to be treated for recycling, an RO unit is installed. About 75% of the effluent fed is recovered as permeate having qualities equal to or better than typical dyehouse soft water, and 25% is rejected to sewer as the concentrate. For the typical effluent solute loads and required permeate quality, the operating pressure of the reverse osmosis unit is around 3 MPa.

2.4 Improved energy efficiency and productivity of the dyehouse

Steam is used to heat water to different temperatures in the dyehouse. When there is no recycle-scheme, the starting temperature for water being heated varies from around 8 °C in winter to perhaps 15 °C in summer, in the case of town water supply. (The range would be wider from a river supply but narrower from a private borehole.)

For the various steps in the process, this water has to be heated to temperatures ranging from 40 to 90 °C, or even 100 °C. The complex operations in the dyehouse produce a mixed effluent that is typically at around 35 °C in the winter to around 40 °C in the summer.

In fact most of the steam consumed in the dyehouse is represented by the difference in temperature between the effluent and the fresh water, i.e. about $35 - 8 = 27$ °C in the winter to about $40 - 15 = 25$ °C in summer.

When a recycle scheme is installed a far smaller quantity of cold fresh water is brought into the dyehouse. Most (perhaps 75%) of the water used is now warm recycled water at say 40 °C. Less steam is needed to heat this water to the temperature of baths above 40 °C, and less heating time is required. Depending on the operations in the dyehouse, some rinses formerly done with cold fresh water will now be carried out with warm recycled water and the effectiveness of the rinse may be improved. Another consequence is that the mixed effluent temperature when using recycle will be higher than before. The change can be calculated from information gathered from a survey of the present operation.

A graphical representation of the dyehouse heat flows before and after implementation of an Archaeus Recycle Scheme is shown in two Sankey diagrams (Figures 2 and 3). In this example the steam consumed in the dyehouse with the recycle scheme was only 46% of the original figure. When the other steam-users in the factory and losses in the system were included, the fuel consumed by the boiler was 70% of the original figure. Clearly, the saving in boiler fuel costs adds significantly to the attractiveness of an effluent recycle project.

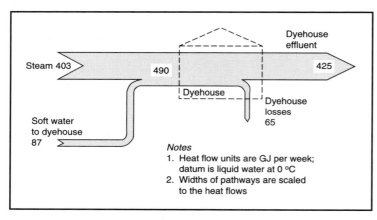

Figure 2 *Heat flows in dyehouse without recycling*

3 Physico-chemical principles

The main processes involved in removal of colour and other contaminants from effluent streams by Arcasorb D are believed to be anion exchange and physical adsorption.

The anion-exchange capacity of the material arises from the presence of electropositive groups, which bind with electronegative ions of anionic dyes. Such a positive group is present on each monomer unit of the polymer backbone. The polymer has two types of functional side-chain, and the method of material preparation affects the proportions of the two types present. The predominant groups are protonated at pH 4 while the second type of group, which is less abundant, is protonated at

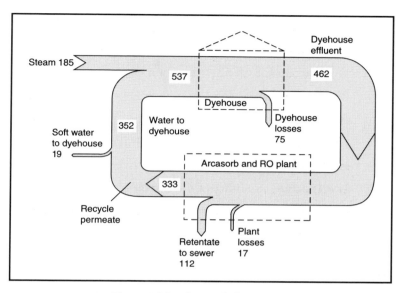

Figure 3 *Heat flows in dyehouse with Archaeus recycle scheme*

pH 6. As a result dye adsorption increases with decreasing pH; detailed testing showed that there is an optimum pH for the raw dye effluent before Arcasorb treatment. Even more dye is removed if the pH is reduced below the optimum, but this has proved unnecessary as proposed NRA targets are easily achieved at higher pH values. Operating the process at higher pH also reduces the amount of acid required and helps to minimise chemical consumption.

Maintaining the system at optimum pH also ensures maximum life of the material, since gradual hydrolysis of the polymer backbone is theoretically possible if the system were maintained at extremely low pH for an extended period. In trials which have been continuing in a demonstration unit for twelve months to date there has been no detectable loss of Arcasorb D performance, and the expected life of the material is in excess of three years in practice.

In addition to ion exchange, Arcasorb D is thought to remove contaminants from waste waters by physical adsorption, that is, the attachment of substances to the surface of the solid. This is in theory distinguished from absorption, in which the substance is distributed throughout the solid, but in practice the two processes are difficult to separate. Since adsorption is a surface phenomenon, more dye or other contaminant will be taken up on material that is finely divided, particularly in the case of large dye molecules; smaller dye molecules are less affected by particle size. Arcasorb D is supplied at the optimum particle size when taking into account clean-up performance and flow characteristics of the bed.

The capacity of physical adsorption processes such as van der Waals forces or hydrogen bonding increases with decreasing temperature in accordance with Le Chatelier's principle, since heat is evolved in the adsorption. This would suggest that the Arcasorb process might perform better at low temperatures but, surprisingly, in laboratory experiments colour removal is better at higher temperatures. This is presumably due to the improved kinetics of the system (i.e. increased mobility of the dye molecules); there is also the possibility that the internal structure of the adsorbent may swell, enabling dyes – in particular, larger dye molecules – to penetrate further into the adsorbent. As a result no temperature balancing has been required on the demonstration unit, where dyehouse effluent temperatures have averaged approximately 40 °C, with peaks of significantly higher temperatures.

Regeneration of Arcasorb D is carried out using a chemical solution made up in raw dyehouse effluent at increased pH. This has several advantages over the use of regenerants made up in fresh water:

- the dyehouse effluent is generally alkaline to begin with and so requires minimal addition of chemicals
- the temperature of the raw effluent is generally 20 °C higher than that of typical town's water, and this assists in the regeneration process but without the requirement for external energy supply
- in addition there is no reliance on a supply of costly town water for this stage of the process.

4 Results of trials

Early laboratory experiments showed that Arcasorb D has a strong affinity for reactive, acid and direct dyes. Single dye solutions were made up to give absorbance readings at the wavelength of maximum absorbance of between 2.5 and 6.25. Colour removal of between 65 and >99% was achieved.

Experimental investigations quickly became concentrated on actual factory effluents, however, since single pure dye salt solutions do not represent the real life problem. Experience has shown that cationic dyes and insoluble dyes such as basic and sulphur dyes are not significantly removed by Arcasorb alone, but may be dealt with by a combination of Arcasorb with other technologies, depending on client requirements. These dyes may not be of major significance in most effluent treatment cases, since insoluble dyes may be removed by filtration while the very high fixation of basic dyes results in low effluent concentrations.

Trials at nine commission dyehouses throughout the UK and Holland have ensured exposure of Arcasorb D to mixtures of all major dye types presented in combination as whole factory effluent. Trials were conducted in a purpose-built demonstration unit comprising an upstream pH-balancing and coarse filtration unit, a single Arcasorb bed (half the length of a full-size bed), full regeneration capabilities and final RO membrane filtration if required for recycle tests. Trials lasted for between one and seven weeks. Typical throughput was $600 \, l \, h^{-1}$ for seven hours per day, with the bed being regenerated every two to three days. True colour measurements were taken every half-hour, and absorbance measurements at 400, 450, 500, 550, 600 and 650 nm were recorded for raw dyehouse effluent and Arcasorb-treated effluent throughout the operation.

For presentation purposes the mean absorbance reading over the six wavelengths was calculated for both the raw and treated effluents to allow graphical description of a full period of operation between regenerations. Figures 4, 5 and 6 show operational results from a 33-day trial of the Arcasorb D demonstration unit at a UK commission dyehouse. Although it is difficult completely to define the dye types present in the effluent at any one time, they would generally consist of 80% reactive, 15% sulphur and 5% basic, metal-complex and disperse dyes.

Figure 4 shows the colour removal perform- ance at each of the six standard wavelengths in a

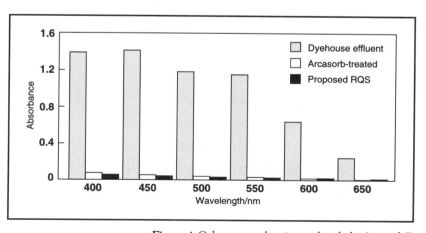

Figure 4 *Colour removal at six wavelengths by Arcasorb D*

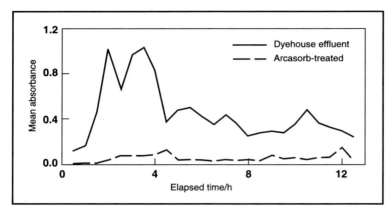

Figure 5 Colour removal performance of Arcasorb D demonstration unit over a single operating period

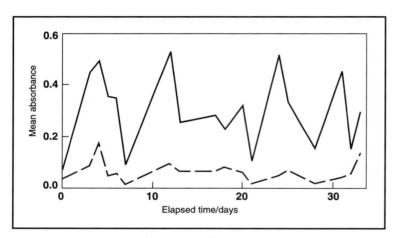

Figure 6 Performance of Arcasorb D demonstration unit: mean true colour removal over 33 dyes (see Figure 5 for key)

spot sample taken on day 24 of the trial. Effective decolorisation was achieved over the full spectrum measured. Since river quality standards were achieved or nearly achieved at each wavelength in this sorbent bed, which is only half the usual size, a full-size bed would clearly surpass this standard easily.

Figure 5 shows the mean colour removal (each point represents the calculated mean of six absorbance values from the six wavelengths) over a full operating period of the demonstration plant. This operating period in fact covered two days of the trial (days 24 and 25) but the sorbent was only regenerated at the end of the second day. Despite a maximum raw effluent colour of 1.033 mean absorbance units, the unit continued to perform effectively, removing 92.5% of the colour. At the end of the operating period colour reduction exceeded 85%.

Dyehouse effluent is inevitably highly variable in colour, pH, temperature, salts loading and so on. A commercial treatment system must be sufficiently robust to withstand such normal fluctuations. In order to gain an impression of the overall daily Arcasorb D colour removal performance, a mean daily average colour was calculated from all the half-hourly measurements across the six wavelengths taken throughout each day for both raw and treated effluents. Figure 6 illustrates the mean daily average performance of the demonstration unit over the 33-day trial period.

Performance targets for Arcasorb D have easily been achieved in terms of colour removal from commission dyehouse effluents containing mixtures of different dye types. In addition to colour removal, levels of chemical oxygen demand (COD) in the treated effluent have also been reduced. Table 1 shows COD reduction in some dyehouse effluents treated on site by the demonstration unit. Results show that the Arcasorb system is a robust and effective treatment process for typical dyehouse effluents, producing decolorised effluent consistently despite highly variable feed.

Table 1 Reduction in COD of dyehouse effluent using Arcasorb D

Dyehouse site	COD of effluent/ppm		Removal /%
	Untreated	Arcasorb-treated	
1	3024	1073	64.5
2	1140	774	32.1
3	831	380	54.3
4	1845	990	46.3

5 Cost-benefit considerations

The use of an adsorbent medium to clean up effluent is neither new (activated carbon has been used for many years on water and gaseous emission clean-up) nor a complete solution. The Arcasorb product has several beneficial features in terms of 'earning its keep', such as removal of soluble colour and COD reduction, but there is still a residual problem of salts and insoluble colour. The extent to which these contaminants are important in a dyehouse effluent will determine the precise process configuration selected.

In addition, the charging formula used by water companies is complex because it depends upon flow rate as well as COD, suspended solids and (in future) colour relative to a standard. Where the standard is well known and reasonably static (as for COD), a treatment strategy can be clearly defined. For colour the means of charging has yet to be laid down and this makes it difficult to define a robust treatment strategy.

Within this uncertainty, three things seem clear:
– colour charges will be imposed in the near future in many parts of the UK
– cost benefit is obtainable by reducing the total load of COD, suspended solids and colour
– cost benefit will also accrue from reduced throughput of water (reduced water, softening, volumetric and system charges).

This points strongly to a recycle system as offering the best combination of cost benefit as well as relative insensitivity to the charging regime.

5.1 Process configuration

The configuration selected will depend upon a combination of factors, including:

— source of water (town or borehole/river), and cost of this water
— cost of sewerage treatment
— contaminants present in the effluent
— funding availability for capital plant
— space limitations
— degree of 'smoothing' in the effluent.

Technically, adsorbent processes are able to withstand rapid and frequent changes of process conditions and therefore the degree of 'smoothing' (or 'balancing') of the effluent stream necessary is very low. Trials of Arcasorb at factories show that 1–2 h balancing time has been sufficient for every location tested to date. This enables quite compact plant designs to be achieved, since a large balancing tank is not required; for example, for a factory producing 1000 m^3 per day of effluent the balancing tank is 4.5 m diameter by 4 m high if built above ground.

Space limitations in dyehouses are frequently severe, but the compact nature of an adsorbent-based system, and the ability to separate parts of the plant, mean that the plant can be located in separate modules around the factory. By way of illustration, the complete treatment plant (excluding balancing tank) for a factory producing 650 m^3 per day is designed within a floor space 11 × 11 m (maximum height 4 m). In general, therefore, constraints of space and balancing do not affect the choice of process configuration for the Arcasorb-based system – it is usually an iterative process between what is technically desirable versus what is a paying proposition.

For a typical commission dyehouse processing a range of fabrics and dye types, and subject to town water and sewerage charges of, for example, £1.20 m^{-3} at present, it is economic to install a recycle system consisting of a combination of Arcasorb, filtration and RO technologies. Where colour charges are not an important factor, or where clean water is available cheap (or free), the installation of an RO unit becomes uneconomic. Some recycle use can, however, be achieved using effluent treated by filtration and Arcasorb. The effluent so treated contains salts and a fine suspension but little colour, so is probably best described as 'grey water'. Its use has been evaluated by several dyehouses. Grey water can be used for rinsing and scouring operations; trials have shown that it could also be used for dyeing some fabrics in certain shades.

5.2 Costs

The costs of owning and operating an adsorbent-based treatment plant arise as follows:
- cost of the plant itself
- replacement cost of active medium, filter materials, RO membranes
- chemical costs, sludge disposal, power
- labour costs
- residual water and sewerage costs.

The capital cost of a full recycle plant using adsorbent technology is high, due to its inherent nature and the level of automation required in high-labour-cost economies. The plant can be built much more cheaply where labour can be relied upon to perform routine tasks such as regenerating an adsorbent bed, and where this is economic. In the UK fully automated plant has been specified, requiring little more than a visual check every few hours and a daily filter clean (a half-hour task).

Conversely, operating costs (in terms of chemicals, sludge disposal and power) are low and in most countries are expected to increase slowly for the foreseeable future (sludge disposal may be the exception). Typical operating figures obtained in trials over 12 months at dyehouses average out at £0.30 m^{-3} effluent treated. This figure comprises all the variable costs of plant operation: chemicals for pH control, bed regeneration and membrane cleaning, sludge disposal to landfill, filter materials and power. Total costs including capital or rental charges and the benefit derived by installing an Arcasorb plant are examined in section 5.4. For nonrecycle plants, operating costs are reduced because of the much reduced power load.

Residual sewerage charges will exist if recycle is less than 100%, and the authors know of nowhere in the world at present where 100% recycle is cost-effective. Due to the concentration effect across an RO membrane it is possible to see an increase in the rate per cubic metre charged for components (e.g. COD) in the effluent, but because the volume has reduced, there is a net saving in the overall costs of sewerage disposal.

5.3 Benefits

After installing a recycle system, money is saved in the following ways:

- reduced water charges
- reduced softening costs
- reduced sewerage charges (volume, COD, suspended solids, colour)
- lower net heat loss from the factory (i.e. energy saving)
- reduced time to heat up dyebaths
- consistently high water quality
- less dependence on water source supplies
- resilience to increases in water and sewerage charges.

5.4 Net costs of treatment system

Table 2 illustrates how the costs and benefits relate for an average-sized dyehouse in the UK, based on real data but converted to a nominal £100 000 per annum water cost in year 1 for commercial reasons. It is assumed that escalation rates per year are as follows: 5% water, chemicals and maintenance, 10% sewerage (colour charge in 1996 onwards) and 4% heat,

Table 2 Relative costs of treatment system

	Without plant				With plant					Overall net benefit
Year	Water	Sewerage	Heat	Total	Water	Sewerage	Heat	Plant costs	Total	
1995	100	79	72	253	29	30	56	98	213	40
1996	105	126	74	305	30	32	58	122	242	63
1997	110	136	77	323	31	32	60	126	249	74
1998	116	145	80	341	33	32	63	129	257	84
1999	121	152	83	356	35	33	65	132	265	91

with plant rental costs fixed for whole period. The stated plant costs include all variable operating costs and rental and maintenance agreement charges. The daily requirements for monitoring of operation and attendance for filter changing and so forth are minimal and can be handled by existing dyehouse personnel. The relative sizes of water, sewerage and overall benefit to the client are believed to be accurate for this type of dyehouse. Future costs are projections which are believed by the dyeing industry in the UK to be conservative.

There is considerable scope with such a system to tailor the cost structure in order to provide a reasonable rate of return to the dyehouse. For dyehouses able to afford the capital cost of purchasing a plant, payback periods of three years and less are easily achieved on UK costings. Plants supplied on a rental basis (including full maintenance of the plant, active medium and membranes) bring immediate increased profitability to a dyehouse.

5.5 Sensitivity analysis and risk factors

The importance to a dyehouse of selecting a good effluent treatment strategy is so great (a small company will need to spend over £500 000 in the next few years, even if they do nothing!) that the robustness of the cost/benefit equation should be considered and tested. There are certain key performance criteria that a treatment plant must achieve in order to generate the savings referred to above (such as recycle quality and operating cost), as well as the operational variability present in any dyehouse, particularly high or low rates of effluent production. This last-named item is not usually a factor in most dyehouse managers' thinking, and it should not become one; however, a treatment facility must be sized on some basis, and the key issue is whether the economics are ruined if the factory should operate below design capacity.

The answer is a reassuring one: the Arcasorb process is relatively insensitive to the most important variables considered both by clients and by the authors. Again, using the 'normalised' figures for our case study, in 1996 the net benefit sensitivities would be as shown in Table 3. The risks of treating effluent at the factory can therefore be seen to be manageable, in terms of the payback from such a project. There are other risks to consider.

Firstly, off-specification recycled water can cause a quality problem in the dyehouse. This risk is, of course, present with town or river water, but is managed by appropriate testing and sampling. With a treatment plant installed the risk of sudden off-specification water is low,

Table 3 Net benefit sensitivities of Arcasorb process

| Item | Variance/% | Sensitivities/$£ \times 10^3$ | | |
		Worse	As designed	Better
Throughput	−20/+10	38	63	76
Operating cost	−20/+20	56	63	71
Heat recovery	−30/+35	58	63	69

due to the inherent nature of membrane technology. Simple and robust condition monitoring can provide automatic alarm and shutdown of the treatment unit.

The second risk relates to plant performance. The suppliers guarantee the plant's performance on delivery, however, and take the risk of long-term deterioration of the active medium and/or membranes through a maintenance agreement. The residual risk of abnormal effluent being put into the system has been discussed above and rests with the dyehouse.

Finally, there is the risk of doing nothing. In a world where resources of pure water are becoming ever scarcer, and where attention to the environment is growing steadily, no management can ignore the impact of disposing of very large quantities of aqueous waste. A cost-effective effluent treatment system can now be seen to be a good business proposition for dyehouses. The consequences of doing nothing will be at least a lost opportunity, and for many it could mean going out of business.

6 Successful installations

Successful pilot trials have been completed at nine dyehouses in the UK and Holland. Extensive data have been collected from these, enabling full-scale plant to be designed. At the time of writing, contracts are being negotiated for the first full-scale ($30 \text{ m}^3 \text{ h}^{-1}$) recycle plant project, with an expected commissioning date of early 1995.

CHAPTER ELEVEN

Treatment of dyehouse effluent for subsequent re-use of reclaimed water

Peter H Weston

1 Introduction

It is now widely accepted by dyehouse managements, on environmental and economic grounds, that they must treat their effluent, and if possible re-use their process water, on a continuous basis. By re-using reclaimed water, it is possible to cover the treatment costs involved (including purchase and operation of the equipment required) within a reasonable period of time, say two to three years. Also, when these costs have been covered, there will be a continuing beneficial effect on the whole economics of the dyehouse operation.

In the Leicester area, the 1994–95 combined water/effluent costs can be averaged out at around £1.10 m^{-3}. In 1992–93, North West Water projected a doubling in the cost of water usage by the year 2000, and nothing that has happened since then has indicated that this projection is inaccurate. Several companies have taken in-house action in order to minimise these seemingly constantly rising costs while remaining compliant.

As an example of the level of costs involved, a dyehouse in Leicestershire taking water at 30 m^3 h^{-1} from the main will pay for 90–95% of that amount to be discharged as effluent. At a metered water rate of £0.70 m^{-3}, this is equivalent to £21 per hour input charge. Allowing for the permissible 5% loss in process allowance given by the water company, the output charge for 95% of this volume at £0.40 m^{-3} will be £11.40 per hour. This gives an hourly combined cost of water plus sewerage of £32.40.

These costs are significant, but without water a dyehouse cannot operate. It is therefore essential that these costs are controlled and minimised. The obvious answer is to reclaim water from the effluent and re-use it continuously. It is relatively easy to identify a system that will work in a particular situation. The questions of cost and payback are equally important, however, and the answers to these are a little more difficult. Indeed, there is no unique answer, and thorough investigations are essential for the identification of a suitable solution for each individual dyehouse.

2 Identifying a system that works

In general terms, it is possible to remove anything from an effluent stream, at a price. But in many cases costs are so high as to be prohibitive. At the time of writing, no commonly available, low-cost method exists for the removal of some contaminants (for example, salt from a reactive dye formulation).

The presence of colour in a dyehouse effluent is currently seen as a particular problem. The extent to which this forms an actual or potential health hazard is a matter that will continue to be debated for many years to come. Yet such a discussion is in one regard beside the point, because the authorities have already decided that colour will become a charging determinand/consent in the very near future.

To install in-house treatment equipment that will clean up the effluent and permit repeated re-use of the water has been the ultimate answer sought for many years. Three years ago Peteson Ltd became involved in this activity, and its first full-scale treatment system was installed after about one year's development work. The installation/commissioning period was not without its difficulties, which were due mainly to the surprisingly wide variety of production requirements in a commission dyeing range. Subsequently a new company, Westover Industries, was formed by the Peteson personnel (and others) in order to provide the textile coloration industry with a BATNEEC (best available technique not entailing excessive cost) answer to the problems that it faces in the context of its water usage.

The solution Westover offers is a system based on a variety of in-line treatment methods, determined according to:
– the profile of the effluent to be treated
– the end-use of the water reclaimed

— the available outlets for removal of the dewatered sludge

— the amount of money available for this purpose.

For the vast majority of cases examined to date (20 dyehouses), a payback period of between 18 months and three years (based on today's costs) has been identified. The imposition of additional charges by the water companies will reduce this period.

The Westover system is based on the premise that it is more efficient and cost-effective to treat a single, combined, effluent in a sequence of the type shown in Figure 1. Whilst each

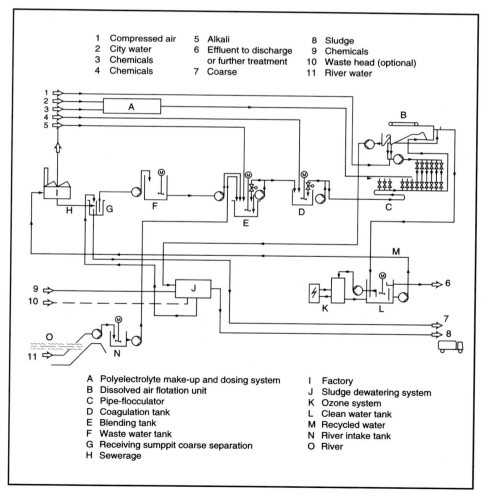

1 Compressed air 5 Alkali 8 Sludge
2 City water 6 Effluent to discharge 9 Chemicals
3 Chemicals or further treatment 10 Waste head (optional)
4 Chemicals 7 Coarse 11 River water

A Polyelectrolyte make-up and dosing system I Factory
B Dissolved air flotation unit J Sludge dewatering system
C Pipe-flocculator K Ozone system
D Coagulation tank L Clean water tank
E Blending tank M Recycled water
F Waste water tank N River intake tank
G Receiving sumppit coarse separation O River
H Sewerage

Figure 1 *The Westover system: flow diagram*

section is designed to perform a particular function, that same treatment will have some effect on other contaminants that require action. These are summarised in section 3 (below), and discussed in greater detail in the following sections.

3 Process outline

3.1 Preliminary mixing

The combined effluents are first mixed together in a single vessel, the main function of which is to equalise and minimise the variations in flow rate and of the quantity and type of contamination present. A relatively high volume of stored effluent provides for better controllability in the subsequent treatment processes.

3.2 pH control

The next stage in the system is the imposition of pH control, as the water reclaimed must be at a pH level that make it suitable for re-use. The identification and modification procedures are carried out in the equalisation vessel.

3.3 Removal of foreign matter

Before the effluent liquor enters the equalisation vessel, all pieces of foreign matter, such as thread, fragments of broken pallet and labels, must be removed.

3.4 Removal of suspended solids

This is the first stage of the Westover process proper. After thorough investigation and cost comparisons, it was decided not to use membranes for this purpose, due to the wide and frequently uncontrollable variation in the particle size and in the overall levels of suspended solids present, and the cost of membrane usage. The preferred method is therefore dissolved air flotation (DAF) (section 4).

3.5 Decolorisation and sterilisation

In the Westover system this important process stage is based on the use of ozone, although hydrogen peroxide can also be used if warranted by specific circumstances (section 5).

3.6 Sludge dewatering

The final stage in the treatment is to dewater the sludge produced so that the water can be re-used (section 6).

4 Dissolved air flotation (DAF)

4.1 Process summary

DAF is a well-known and proven method of removing suspended solids from water. In the process air is added to an effluent flow, often in the presence of chemicals (coagulants and flocculants, for example) to facilitate the separation of the solids from the water.

Effluent, chemically modified or otherwise, is pumped into a flotation unit at the same time as air is dosed into the suction side of an aeration pump. This air/water mixture is pressurised in the pump, and the air is dissolved in the water. Subsequent depressurisation results in the formation of microscopic bubbles which adhere to particles in the water, giving them the buoyancy they require to rise quickly to the surface. The rate at which the particles rise can be determined by applying Stokes' law, and this information, combined with the known flow rate, identifies the separation area required.

It is possible to improve the solids removal by adding appropriate chemicals. In many industrial situations the following reductions have been achieved on an on-going basis:

– suspended solids removal of 99% or more
– COD reduction of 90%
– fats/grease removal of 98%.

Once the solids have risen to the surface of the liquor in the DAF unit, they are skimmed off and transferred to a sludge reservoir prior to further treatment in the dewatering phase. The larger particles that could not be 'floated out' on the surface settle in the DAF tank (via a series of plate separators) and are pumped to the same dewatering stage as the skimmed solids.

The remaining 'dirty' water, free of solids, is then pumped to the next stage of the treatment process for removal of colour and bacteria. In the Westover system (Figure 2) this is achieved by pumping through an oxidising section (see section 5).

Figure 2 Dissolved air flotation unit

4.2 Chemicals used in a typical DAF process

Coagulant

Coagulation is the term given to the destabilisation of the colloidal particles in an aqueous suspension. Such particles are less than 1 mm in size and tend to repel each other due to their intrinsic surface charges. Coagulants neutralise such charges, allowing the particles to approach each other. During subsequent flocculation, the particles accumulate into clusters.

Flocculant

A flocculant is a chemical reagent that will aid the separation of solid particles from aqueous media by the agglomeration of such particles into larger clusters (flocs). These will settle, float, filter or separate more easily, and can be removed by liquid/solid separation techniques such as DAF. The mechanism of flocculation is complex, and can briefly be described as a combination of a change in the ionic balance of the particles, and their adsorption on to the molecular chain of the flocculant. There is, however, no universal flocculant that can cater for all conditions.

4.3 Dosing of flocculants into an effluent stream

Flocculants are high-r.m.m. products, the molecular chains of which rupture under high shear conditions. Violent mixing or pumping of flocculating mixtures must therefore be avoided, or efficiency will be irretrievably impaired. The solid particles to be treated rapidly adsorb the flocculants; to ensure efficient flocculation, it is necessary that particles and flocculant are brought into close contact. For good technical and commercial efficiency, addition of flocculants should be:

– from very dilute solutions
– as near as possible to the point where flocculation is required
– at areas of local turbulence, and if possible at different points within this area
– across the whole stream of liquor to be treated
– to a region where excessive turbulence, after floc formation, is avoided.

4.4 Dose rates

The quantity of flocculant necessary to obtain a good solid/liquid separation is very small, with typical figures for different systems being:

(a) high-solids-content mineral systems 2–20 g m^{-3}
(b) low-solids-content mineral systems 0.5–2.0 g m^{-3}
(c) low-solids-content organic systems 0.25–2.0 g m^{-3}.

5 Decolorisation and sterilisation

5.1 Ozone processes

The effluent liquor from a dyeing process contains dyes, textile auxiliaries and salts. Advanced techniques for using combinations of ozone and hydrogen peroxide make it possible to remove odour, colour, chemical oxygen demand (COD), total organic carbon (TOC) and adsorbable organohalogens (AOX) (Figure 3). The resultant products are often easily biodegradable. Ozonation at higher pH and irradiation in the presence of hydrogen peroxide both lead to the production of hydroxyl radicals, which can react with organic material.

The behaviour of dyes, and their reaction with hydroxyl radicals under different conditions, have been studied in both the presence and the absence of auxiliaries. Both ozone and hydrogen peroxide, individually and in combination, have proved effective in decolorisation. However, to achieve the best balance of desirable properties in the continually varying working conditions of a textile dyehouse, they need to be present at carefully

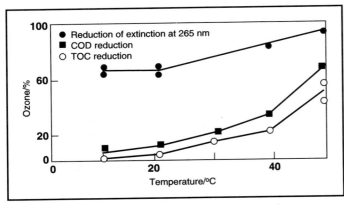

Figure 3 *Variation of the effectiveness of ozone treatment with temperature*

determined concentrations. These optimum conditions can be identified only by carrying out in-house testing over a prolonged period.

Westover Industries concentrates on the use of ozone as the primary decolorisation method. Peroxide is sometimes employed to improve process efficiency or to reduce costs.

Ozone is being increasingly used for the disinfection, decolorisation and deodorisation of potable and other waters, for purifying water intakes into power and desalination plants, and for the production of ultrapure water for the semiconductor and pharmaceutical industries. It is gradually replacing chlorine for these purposes (in spite of being more expensive) owing to the undesirable nature of chlorine residues in the environment. Ozone (O_3) is produced industrially by passing an electrical current through oxygen gas. When the ozone is added to the solids-free 'dirty' water its powerful bleaching effect removes the colour and the bacteria from the water, efficiently and at reasonable cost.

Ozone is a strong and rapidly acting oxidising agent, and when properly applied will destroy all bacteria and viruses. It removes by oxidation the organic by-products of dead bacteria in treated water, and as its active life is short, the necessary safety provisions can easily be built into an on-line industrial unit.

Even so, great care is needed when handling and using ozone (Figure 4). The systems supplied by Westover have several built-in safety features designed to prevent potentially hazardous ozone escapes. Any ozone leakage in the equipment immediately triggers shutdown of ozone production. The ozone already produced in the system is held under reduced pressure until it can come into contact with incoming oxygen, with which it reacts to produce more oxygen. Any residual ozone is passed to a manganese converter, within which it is rendered harmless.

Overproduction of ozone is costly, and any shortcomings in the treatment systems prior to ozone treatment will mean reductions in efficiency and increased costs. The temperature must be maintained at the correct level and pH must be properly controlled.

Research into improving the efficiency of these processes is continuing. Many promising developments employ treatment under ultraviolet radiation.

5.2 Salt

It is essential that the question of salt usage is examined as part of the initial investigation, so that the necessary removal requirements can be identified and included in the system. Dissolved

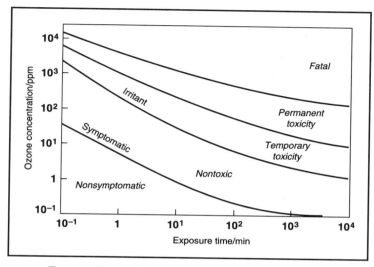

Figure 4 *Human tolerance of ozone, showing different regions of toxicity*

salt (sodium chloride) will not be removed by ozone treatment, but salt removal need not be a problem technically. Many effluents do not contain salt, and sometimes it is not necessary to remove all the salt from the water reclaimed for re-use. Moreover, salt normally represents only a small proportion of the final dyehouse effluent, which includes liquors from washing, scouring, fixing and other processes. Typically salt is present at levels between 0.5 and 2.5% of the total effluent.

Where the presence of salt does cause problems, then complete removal is possible by modifying the system described, but at additional cost. If done properly the resultant reclaimed products, clean salt (brine) and water (and possibly other by-products) may be usable in the dyehouse.

6 Dewatering

The final, but important, stage in designing industrial effluent treatment is deciding what to do with the solids (sludge) removed from the effluent. Westover recommends dewatering of the sludge and return of the water for re-use. There are many methods of sludge dewatering, and the main factor to consider is cost. It is not currently necessary (or usual) to have to dry the sludge to above 35–40% solids content for landfill discharge.

Decanters work well (better with the addition of a flocculant) and achieve a dry-solids content of over 35% easily and continuously. They need little attention other than routine observation and maintenance. Some are expensive, but second-hand systems can be purchased at reasonable prices.

Filter presses are either manual or automatic in operation. Manual ones are inexpensive, but of course require the constant presence of an operator. Automatic filter presses are efficient but costly. Both types can achieve a high proportion of dry solids in normal operation.

Vacuum filters (belts) that have been examined by Westover Industries do not appear to be the best option, but that could be the fault of the particular equipment tested (the work continues).

In the foreseeable future it is likely that landfill for the discharge of solid industrial waste will not be available, or will be very expensive, so alternatives must be identified and investigated.

7 Quantifying the savings

7.1 Background

An essential prerequisite to any discussion of the costs of effluent treatment systems is a realisation that dyehouses do not, in the 1990s, have any option on whether or not to clean up the waste they produce. Coupled with this is the rising cost of water, which is used in great quantities in all dyehouses. So the basic requirements in every case are to control and minimise costs in order for individual dyehouses to remain competitive, whilst providing the necessary level of environmental protection.

Peteson Ltd became involved in this issue following the appearance of press reports of a possible loss of 40 000 jobs in Leicestershire alone. The company's interest in the problem was a natural development, as it was already concerned with the treatment for re-use of effluent waters in many other industries throughout Europe. Its emphasis is always on the need to achieve economical treatment of aqueous dyehouse effluents so as to enable the reclaimed waters to be re-used continuously.

7.2 Savings identified

Every dyehouse can identify the volume and cost of the water it receives: it is the sole basis for its water/effluent bills. For the purpose of the following example, a water input rate of $30 \, m^3 \, h^{-1}$ has been assumed.

Water is lost in many parts of the dyehouse. Most is steam loss to the atmosphere, but also there are uncontrollable discharges to surfaces, leakages and washdowns, as well as water used in kitchens, toilets and washrooms. The sewage undertaker (who is often also the water supply company) normally allows for losses in the process area, the norm being only 5%. So on this basis the dyeworks requires a treatment system that can handle 95% of the intake water volume, i.e. $28.5 \, m^3 \, h^{-1}$. In practice, however, such a system would be far too large, and if installed would provide overcapacity for any future increases in production.

Using the $28.5 \, m^3 \, h^{-1}$ figure, and in the light of Westover experience in dealing with suspended solids contents of 0.5–5.0%, a typical figure for suspended solids content would be about 2%, which gives $28.5 \times 0.02 = 0.57 \, m^3 \, h^{-1}$ throughput of suspended solids. This throughput can be used to calculate the amount of water and chemical floc that must be present in order to carry the solids to the dewatering unit. If we assume that there is a solids content increase to $1.20 \, m^3 \, h^{-1}$ in a total sludge flow of $3 \, m^3 \, h^{-1}$ sent to be dewatered (work is in progress to enable more precise estimates to be made in this area, this will result in the figures shown in Table 1 for the amount of water separated. Using this total figure gives the savings quoted in Table 2.

Table 1 Amount of water removed in a typical Westover system (see text)

Water separated	Amount/$m^3 \, h^{-1}$
In the DAF	25.5
In dewatering	1.8
Total	27.3

Table 2 Typical hourly savings

Item	Amount of water removed /$m^3 \, h^{-1}$	Charge /£ m^{-3}	Saving /£ h^{-1}
Water input	27.3	0.70	19.11
Effluent output	27.3	0.40	10.92
Total			30.03

Our typical dyehouse works 24 h per day, 5.5 days per week, 48 weeks per year. So the annual saving could be £190 270, to set against the cost of the equipment required to treat that volume of effluent (including in-house modifications) – in the region of £250 000. The estimated cost of operating the equipment is around 25% of the purchase price, i.e. £62 500. This gives a total cost over a two-year period of £375 000. With annual savings of £190 270, there is thus a payback period of just under two years. Naturally savings will continue after the end of the second year, probably in line with the increasing costs of buying water and discharging effluent.

CHAPTER TWELVE

Membrane technology for the separation of dyehouse effluent

Clifford Crossley

1 Introduction

This paper provides a description of the different membrane processes, including an outline of the relevant theory in the fields of reverse osmosis (RO) and nanofiltration (NF); the variations in theory for ultrafiltration (UF) are added on a more general basis. This is followed by a discussion of the various membrane configurations and of their merits and limitations. Typical applications for different membrane processes are presented, together with a general discussion of the way membrane systems are structured. The opportunities for the use of membrane processes for the treatment of textile effluent are discussed, and details are presented of the work carried out and the typical systems proposed.

The costs of water and its disposal have increased considerably over the past few years, and at the same time discharge limits for effluent disposal have been tightened. Some of the membrane applications mentioned in this paper may not be economically feasible at present in all cases, but they may become so in the near future, because of increasing water costs.

2 Theory of membrane separations

2.1 General principles

The osmotic pressure P_{osm} exerted by an ideal solution is given by Eqn 1:

$$P_{osm} = \frac{nRT}{V} \tag{1}$$

where n/V is concentration, R the universal gas constant and T the absolute temperature.

The solvent (water) flux through a semipermeable layer is given by Eqn 2:

$$J = A(\Delta P - \Delta P_{osm}) \tag{2}$$

where J is flux ($1\ m^{-2}\ h^{-1}$), ΔP the trans-membrane pressure, ΔP_{osm} the osmotic pressure differential between feed and permeate, and A is a constant.

Eqn 2 shows that the flux is limited by the osmotic pressure and thus the solute concentration, as the driving pressure needed in order to achieve a reasonable flux normally is at least 1.5–2.0 MPa. The osmotic pressure of a sodium chloride solution at various concentrations is shown in Table 1, as an example. The limiting concentrations attainable using standard reverse osmosis for a range of typical solutions are shown in Table 2.

Table 1 Osmotic pressures of sodium chloride solutions

Concentration /mg l^{-1}	Osmotic pressure /MPa
1500	0.12
5000	0.39
12000	0.93
35000	2.78
50000	4.04

Table 2 Maximum concentrations attainable using standard reverse osmosis

Solution	Concentration/%
Sodium chloride	5
Sodium sulphate	10
Protein	25
Sugar	30

For a membrane system (reverse osmosis and nanofiltration) the working equation for the flux as a function of concentration has the form of Eqn 3:

$$J = k_1 - k_2 C \tag{3}$$

where k_1 and k_2 are constants, and C the solute concentration.

The solute passage S is calculated from Eqn 4:

$$S = \frac{C_P}{C_b} \tag{4}$$

where C_p and C_b are permeate and bulk concentrations respectively.

For ultrafiltration the working equation has the same form, but the concentration term in the flux equation is generally a natural logarithmic function.

In general, membrane performance is characterised by the flux and the passage of species through the membrane surface. In the case of ultrafiltration and microfiltration (MF), the passage of the species in the permeate will depend on molecular mass or on actual particle size. For reverse osmosis and nanofiltration, passages are quoted on the basis of the ionic species that are retained or passed as a percentage of the total stream concentration. The terms *salt passage* and *rejection/retention* are used for these membranes. The principal differences between reverse osmosis, nanofiltration and ultrafiltration are summarised in Table 3.

Table 3 Principal differences between reverse osmosis, nanofiltration and ultrafitration

System	Pressure /MPa	Crossflow rate /m s^{-1}	Process flux/l m^{-2} h^{-1}	Retention
Reverse osmosis	3–6	2–3	5–40	>90% NaCl
Nanofiltration	2–4	2–3	20–80	>90% lactose <50% NaCl
Ultrafiltration	0.5–2.5	3–4	5–200	4–200 × 10^3 MWCO[a]

a See text below

2.2 Reverse osmosis

In reverse osmosis, a pressure greater than the osmotic pressure of the solution is applied to the solution while it is in contact with a semipermeable layer (membrane). Water (permeate) flows through the membrane surface (permeation), and the solution retained by the membrane is concentrated.

2.3 Nanofiltration

Nanofiltration is very similar to reverse osmosis and is sometimes called 'loose reverse osmosis', as the pressures of operation are often similar. Nanofiltration membranes, however, unlike those for reverse osmosis, allow the passage of various amounts of ionic species. These are mainly monovalent salts, or small amounts of divalent salts. The membranes generally have molecular weight cut-offs (MWCO) in the range 200–300.

2.4 Ultrafiltration

Ultrafiltration is a pressure-driven membrane process, but with pressures in the range 0.5–2.5 MPa. The flux performance of the ultrafiltration membrane is also determined by the crossflow velocity to a greater degree than in reverse osmosis or nanofiltration.

2.5 Microfiltration

Microfiltration separations require very low pressures, typically between 0.1 and 0.5 MPa, and the separation performance is greatly affected by the crossflow rate and the trans-membrane pressure. Typical fluxes expected in microfiltration are in the range 80–250 l m^{-2} h^{-1}, and may be even higher. Microfiltration is generally used to separate macromolecular and suspended material.

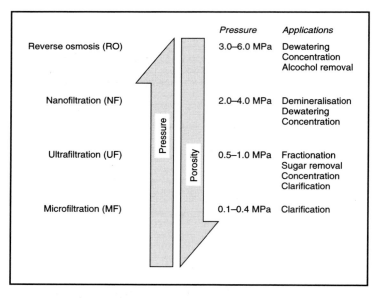

Figure 1 *Membrane technologies: pressures used*

The differences in the pressures used in the four processes and in the molecular sizes retained are shown diagrammatically in Figures 1 and 2.

3 Advantages of membrane separation processes

The main advantage of a membrane separation process is that concentration is achieved without any input of thermal energy, or a change of state, making the process energy-efficient. Membranes provide a finite separation barrier, which allows very dilute solutions to be concentrated and separated. High recoveries can often be achieved, and in some cases valuable product can be recovered from a dilute waste stream.

Membrane systems, whether organic (polymeric), ceramic, reverse osmosis, nano-filtration, ultrafiltration or microfiltration, offer a range of unique separation capabilities in many different configurations, and materials, allowing a very wide variety of process applications.

Compared with other separation or concentration systems, their space requirements are low, and their modular construction and design allows relatively easy expansion.

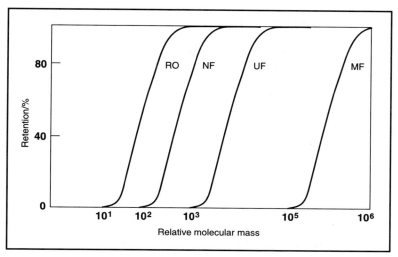

Figure 2 Membrane technologies: molecular sizes retained

4 Limitations of membrane systems

Membrane systems are limited in several ways, as a consequence of their chemical structure. The use of organic or polymeric membranes is generally limited to temperatures below 70 °C and pressures of less than 6–7 MPa, with a pH range of 2–12 (maximum), depending on the type of membrane and its chemistry. Ceramic membranes (ultrafiltration and microfiltration) can withstand higher temperatures and extremes of the pH scale, but are limited to low-pressure operation.

Polymeric reverse osmosis membranes compact with continued use, and this leads to a slow decline in flux with time; the rate of decline will depend on the process fluid, the operation conditions and the cleaning regime used. This decline, together with fouling and caustic attack from cleaning, will determine the life of the membranes.

The biggest limitation of a membrane system, as opposed to a heat-driven concentration process, is the concentration achievable. Membrane concentration (reverse osmosis) is limited by the osmotic pressure exerted by the process fluid. Table 1 shows the osmotic pressure of salt, at different concentrations, and Table 2 gives examples of the limiting concentrations of typical solutions using reverse osmosis. The development of polymeric membrane systems capable of withstanding high temperatures (up to 90 °C) and high pressures (up to 15 MPa) is under way at present, however, and these developments can be expected to push back the concentration limits of reverse osmosis systems.

5 Membrane configurations

The main membrane configurations for organic and polymeric membranes are usually supplied in either tubular, spirally wound, flat-sheet or hollow-fibre form. Ceramic membranes are normally made in tubular or multichannel configurations, but have recently also been engineered in flat-sheet and spiral forms. More recently membrane configurations have included spinning-disk and vibrating systems, which are variations of the flat-sheet system. A summary of the general merits and limitations of the main membrane configurations is given in Table 4.

The main difference between the tubular format and the spiral, hollow-fibre and flat-sheet systems is the channel height, which limits the amount of suspended material and the viscosity of the stream which is being separated. Diameters of the tubular systems are generally between 12 and 25 mm, and these can handle high-viscosity streams containing large amounts of suspended material.

A tubular system consists of several modules, each of which contains a bundle of perforated stainless steel tubes, typically 4 m long, each lined by a membrane (Figure 3). The tubes within the bundle are linked by an end-cap arrangement, which is varied for different

Table 4 Summary of merits and limitations of different membrane configurations

Spiral	Tubular	
Laminar flow	High Reynolds number	
Low power per unit area	High tolerance to	
Low cost per unit area	suspended material	
High packing density	Low packing density	
No tolerance to suspended	High power per unit area	
material	High cost per unit area	

Flat sheet	Hollow fibre	Ceramic (tubular)
Laminar flow	Low tolerance to	Excellent thermal stability
Low power per unit area	suspended material	Available in UF and MF
No tolerance to	Low Reynolds number	High cost per unit area
suspended material	Medium packing density	High resistance to solvents,
High packing density		oxidants
Low cost per unit area	Medium power per unit area	High tolerance to suspended
		material

applications, to allow for the processing of highly viscous liquids. The tube bundle is contained in a shroud which collects the permeate, and this shroud has ports to allow the removal of the permeate.

The spiral system has a membrane surface in the form of envelopes, which are separated by mesh spacers

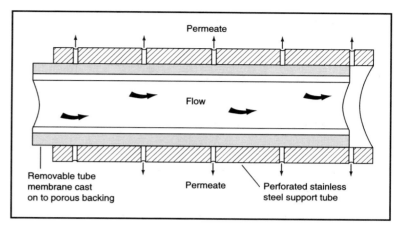

Figure 3 PCI tubular system

(Figure 4); in some systems the spacers vary in thickness. The permeate, once through the membrane surface, passes to the permeate tube at the centre of the element, where it is collected and removed via the end-cap. The end-cap assembly separates the feed, permeate and concentrate streams. Spiral elements are constructed in various diameters (typically 6.3, 10.0, 20.0 and 27.5 cm), and are generally 1 m long. Within a system, up to six elements can be placed in series within a pressure vessel; the system will consist of several pressure vessels, possibly in series, and several stages.

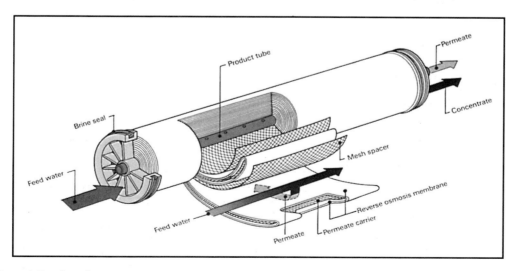

Figure 4 Spiral membrane system

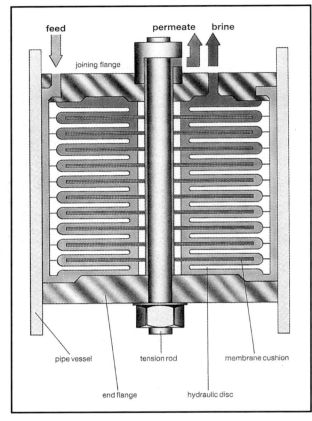

Figure 5 *Plate-and-frame system*

The flat-sheet or plate-and-frame system is similar to the spiral system in that the membrane surfaces are separated by spacer plates, and that the permeate, once through the surface, travels to the centre of the system (sometimes to the edge), where it is collected and removed (Figure 5). Modules for flat-sheet systems are built up as a stack of plates, and these are linked in parallel within a stage, or in series from stage to stage. The channel height is typically less than 1 mm, and the flow is laminar.

The hollow-fibre membranes vary in diameter from 1–2 mm to that of a human hair. Hollow-fibre modules comprise bundles of fibres which are packed in a shroud, sealed in epoxy resin at each end; Figure 6 shows an individual fibre in section. The flow is from the shroud (around the outside) to the centres of the individual fibres and out at the ends of the module. Where the stream contains some suspended matter, the unit is operated in a back-pulsed manner to remove any deposits from the surface and then is removed from the system. Some hollow-fibre systems operate from the inside to the outside; the basic principles are the same.

6 Membrane applications and system designs

Typical applications for membrane processes (reverse osmosis, nanofiltration, ultrafiltration and microfiltration), are included in Table 5. The number of effluent/waste water applications is growing rapidly with the need to re-use water and reduce disposal volumes.

Both continuously operating and batch systems are available, with each of these categories having several variations: continuous multiple stage recycle (MSR), continuous tapered, batch recycle, topped batch and feed-and-bleed are all produced. Schematic diagrams of these membrane system configurations are shown in Figures 7 and 8.

Continuous tapered systems were originally developed for water desalination applications, but were also used in the earlier tubular membrane systems. They are still used extensively in desalination plant

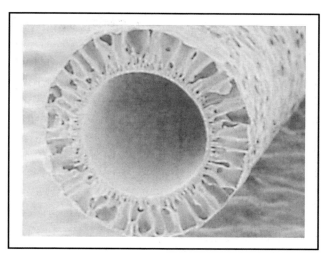

Figure 6 Hollow-fibre membrane

with spiral membranes. These systems require constant feed flow and feed composition in order to maintain stable operation.

Multiple stage recycle (MSR) designs for reverse osmosis, nanofiltration and ultrafiltration are commonly used for waste water treatment, as well as in food, beverage, chemical and pharmaceutical applications. Most membrane configurations (spiral, tubular,

---

Table 5 Typical applications for membrane processes

Reverse osmosis (RO)	Nanofiltration (NF)
Fruit juice concentration	Concentrating and desalting dyes
Desalination of seawater	Concentrating and demineralising whey
Whey concentration	Concentration and desalting of chemicals
Skim milk concentration	Softening of water
Landfill leachate concentration	
Chemical effluent concentration	
Colour/turbidity removal in potable water	
Pesticide and herbicide concentration	

Ultrafiltration (UF)	Microfiltration (MF)
Bleach liquor effluent separation/concentration	Bacterial removal
Oil/water emulsion separation/concentration	Fruit juice clarification
Clarification of fruit juices	Biomass separation
Separation/concentration of enzymes, antibiotics	Pigment and latex separations
Wool scouring effluents	

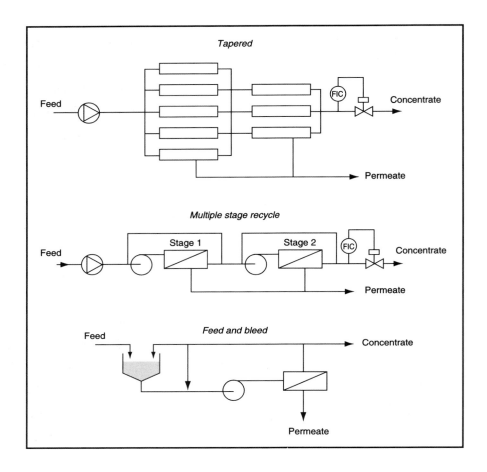

Figure 7 *Typical continuous plant configurations*

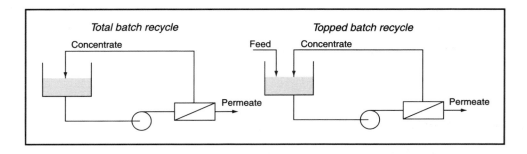

Figure 8 *Typical batch plant configurations*

flat-sheet, etc.) can be incorporated, and these systems are used where the feed concentration varies and to achieve high concentration factors in a continuous operation. The MSR systems normally have higher capital costs then the tapered and the feed-and-bleed systems.

The use of batch designs and topped batch systems is more common for ultrafiltration and microfiltration applications, such as broth or juice clarifications and biomass separations. They are simpler to operate than other systems, and their costs are lower.

7 Applications of membranes in the textile industry

7.1 Wool scour effluent concentration

The separation and concentration of wool scour effluent, using ultrafiltration after the removal of the lanolin and other recoverable constituents, is well known. With the use of 25 000 MWCO membranes, overall recoveries of 75% are achievable. A bulk permeate quality of less than 25 000 mg l^{-1} COD is achieved from a feed of 150 000 mg l^{-1}. Yarn scour effluents are processed in a similar manner, but tighter membranes are used and the feed COD is lower, typically 40 000 to 60 000 mg l^{-1}. Reference plants exist for these applications. The problem still outstanding is the disposal of the final concentrate, which in most cases contains traces of pesticides. It may be possible to use the recovered permeate directly, or it may require further polishing, using reverse osmosis or nanofiltration, before re-use.

7.2 Scouring liquor separation and re-use

On some plants fabric scouring is a stand-alone process, which allows the use of a stand-alone membrane system for the separation of the starches, knitting oils and other contaminants from this stream. Ultrafiltration is the required process, since the pollutants are large molecules. Ceramic membranes may be used, as they allow high-temperature operation with minimal heat loss from the liquor that is re-used.

7.3 Individual dyebath separation and concentration

Where nonreactive dyes are used, dyes can be recovered from individual dyebaths by using ultrafiltration and nanofiltration; this is possible because of the difference in size between the nonreactive and the reactive dye molecules. This application has been successfully carried out in the USA using tubular (polymeric) and ceramic membrane systems.

The separation of colour from dyebaths is technically feasible using nanofiltration. This process may require pH adjustment, in order to suit the membrane operating specification. This process application is very similar to the desalting of product dyes (reactive dyes), for which membranes are widely used. Tubular membrane systems are extensively used for the desalting application because of their wide flow paths and ability to handle viscous liquors.

Because nanofiltration membranes are able to pass monovalent cations while retaining divalent cations, it may be possible to use the technique to recover sodium chloride solutions from dyebaths containing reactive dyes, for re-use.

7.4 Softening of incoming process water

The incoming process water in most cases needs to be treated or softened, depending on the source. This softening or treatment varies from area to area; where the water is very hard or contains turbidity and colour, membrane processes are available for this separation. The Fyne process, using nanofiltration membranes in tubular or spiral form, is being successfully used in Scotland for the treatment of humic or coloured water for potable use.

7.5 Bulk dyehouse effluent treatment

The treatment of bulk dyehouse effluent with an end-of-pipe system would require a variety of different technologies to achieve a high level of recovery for re-use, together with a concentrate discharge quality that meets the consent levels and which is economically feasible. A reverse osmosis membrane system within the overall treatment system would provide the means to recover water for re-use and thus be the main contributor to the economic feasibility, should the overall system be feasible. The bulk of the trials work mentioned in this paper relates to the removal of colour from effluent streams of this kind. Reverse osmosis membranes with, typically, a 95% salt rejection would produce a permeate (recovered water) that would require no further softening before re-use. Moreover, the maintaining of the bulk effluent temperature would allow for the recycling of water that is above the town/river feed temperature, and thus provide low-grade heat recovery. The use of the spiral membrane configuration will offer the best opportunity for an economically feasible process for this application.

7.6 Recovery of treated dyehouse effluent

Treated effluent from the direct discharge site may offer the possibility of recovery and re-use. This may be particularly relevant where the water is obtained from a groundwater source and where the abstraction limits that have been set may be exceeded with increased production. Membrane polishing using a spiral nanofiltration or ultrafiltration system may be sufficient for this application.

8 Payback parameters

The important parameters in the evaluation of textile and dyehouse effluent streams are:
- flux performance with concentration
- maximum concentrations or water recovery
- pretreatment requirements
- operating costs, which include membrane replacement costs, power requirements and chemical requirements for cleaning.

The flux performance with concentration will determine the membrane area requirements and thus the plant size. The selection of the membrane configuration required for the application will determine the feasibility of the process, as the cost per unit area of the different membrane configurations varies greatly. For effluent applications, the use of spiral membranes may be the only option to achieve a economically feasible process where:
- the flows are large
- the suspended matter can be removed
- no viscosity problems will be experienced
- high recoveries are required.

In the separation and treatment of dyehouse effluent streams and individual dyebath liquors, the salt concentration may limit the maximum concentration achievable and thus the systems recovery. The use of a two-stage process (nanofiltration and reverse osmosis), if feasible, would allow higher systems recoveries.

Pretreatment requirements need to be carefully considered for the system that has been selected. These requirements vary from site to site and for different process streams, as well as for different membrane configurations. When considering the bulk effluent stream, general pH and temperature buffering would be required in order to ensure a stable feed stream to the membrane unit and to achieve stable plant operation.

The removal of suspended material such as lint, fibre and other solids is required for the spiral and flat-sheet membrane formats; in general the feed needs to have an SDI (silt density index) of less than 5, with screening down to 10 μm. For tubular membrane formats such a high degree of screening is not necessary, and usually screening to 50–200 μm is carried out. The removal of suspended solids, which contributes to the overall discharge costs, can be achieved by a two- or three-stage screening process; this could include parabolic screens followed by bag filters and even cartridge filters.

The power requirements and membrane replacement costs will represent the major portion of the operating costs. The choice of the spiral membrane configuration offers the lowest power consumption and cost per unit area, and thus the best opportunity to achieve a viable system with a reasonable payback period.

The manning requirements of a membrane system depend on the level of automation. Typically this cost might represent about 20% of the overall operating costs, and the cost of cleaning chemicals perhaps 10–15%.

9 Works trials

An extensive amount of trials work has been carried out by PCI at various dyeing and textile industry sites over several years. The most recent work was carried out at Westertex (Loughborough) Ltd, and centred on colour removal from the bulk effluent stream. Other associated effluent streams, namely scouring liquors and bulk effluent (already decolorised), were also tested. Tubular, spiral and ceramic membrane formats have been tested, and the processes and combinations examined have included reverse osmosis, nanofiltration and ultrafiltration.

The results showed that the use of ultrafiltration is not suitable for reactive dyes, as colour passes to the permeate. The dyes used all had r.m.m.s of 500–2000; the membranes tested had MWCOs in the region of 4000. No examination was made of the types of dye that were retained by the membranes or appeared in the effluent. The use of a nanofiltration and reverse osmosis process combination was extensively tested, but this seems unlikely to be feasible.

Most of the initial work was carried out on tubular membranes (using nanofiltration, reverse osmosis and combinations of the two), and results from these indicated that the use of the tubular membrane configuration would not be economically feasible for the bulk effluent stream. Subsequent and more extensive trials were carried out on the use of spiral membranes using reverse osmosis alone and the nanofiltration plus reverse osmosis combination. The results showed that with pretreatment the use of spiral reverse osmosis alone on bulk dyehouse effluent was possible, with fluxes in the range of 30–$10 \, \mathrm{l \, m^{-2} \, h^{-1}}$. The choice of method for colour removal or destruction, and its associated costs, would determine the overall economic feasibility of the process. In some cases the maximum recovery achievable from the reverse osmosis plant might be limited to 75–80%, depending on the average salt concentration of the bulk effluent stream, which was in the range 4–$6 \, \mathrm{g \, l^{-1}}$.

The fabric scouring trials showed that polymeric and ceramic ultrafiltration membranes with a MWCO between 10 000 and 25 000 are suitable for the removal of most of the starches and knitting oils, assumed to represent approximately 80% of the COD.

The concentration and recovery of water that has been decolorised using an adsorption process was found to be a good application for reverse osmosis using spiral membranes. The fluxes were, however, found to have been lowered by the pH drop required for the decolorisation. Up to 75% of water could be recovered for re-use.

10 Proposed systems

Based on the trials work results and design studies, the systems proposed have included:
- multiple stage and tapered reverse osmosis using spiral membranes for bulk effluent and decolorised bulk effluent
- batch and topped batch systems for scouring liquors and individual dyebath liquors.

Systems that could be used for groundwater treatment, namely water softening or colour/ turbidity removal, would include spiral or tubular nanofiltration membrane in a tapered system design. The polishing of treated effluent for re-use or the concentration of decolorised bulk effluent would require the use of a continuous spiral reverse osmosis plant to achieve economic feasibility.

11 Conclusions

Membrane systems in different configurations do have applications in the dyeing sector of the textile industry in providing a means to recover water for total treatment systems or as stand-alone processes for scouring or dyebath liquor separations. The economic feasibility of some of the processes discussed in this paper depend on the site, but increased water and effluent charges, tighter consents on effluent discharges and new developments and improvements in membrane technologies, are likely to lead to the increased use of membrane processes in the future.

CHAPTER THIRTEEN

Colour removal from dyehouse effluents using synthetic organic coagulants

Peter J Hoyle

1 Introduction

The removal of colour from dyehouse effluent using synthetic organic polymer coagulants represents one aspect of widespread existing technology aimed at the precipitation of water-soluble substances using chemicals of this kind. Indeed, the Allied Colloids product range for residual dye precipitation was born out of an earlier requirement for the analogous treatment of upland, peat-stained raw waters for use as drinking water. The use of these polymer coagulants, typically low-r.m.m., highly charged cationic polyelectrolytes, greatly reduced and often completely eliminated the need for the large amounts of aluminium- or iron-based salts that had traditionally been used.

In the field of colour removal from dyehouse effluents, no such effective and simple inorganic option existed for many soluble dyes. It was, however, recognised that variations on existing polymer coagulant products might allow simple treatment of these previously problematic highly soluble dye residuals to the desired standards. The addition of a suitable polymer coagulant promotes a precipitation of the dye residues, forming small insoluble coloured particles and leaving a correspondingly reduced level of colour in the liquor. The greater the efficiency and suitability of treatment conditions, the greater is the degree of dye precipitation and hence of colour removal. The resulting small colour particles may then be removed from the liquor using a suitable solid/liquid separation process.

A typical process might involve agglomeration of the small colour particles so produced by the use of high-r.m.m. organic flocculants forming readily separable 'flocs' which can be removed from the liquor – for example, via a dissolved air flotation process. This paper describes the general principles behind this technique, its capabilities, the equipment required, the associated costs and other relevant issues.

2 Background

Before discussing the details of the polymer coagulation process for colour removal, it is well worth examining the challenge facing this technique. The National Rivers Authority (NRA) has defined the residual colours present in dyehouse effluent in terms of their ability to absorb monochromatic light at specified wavelengths. For example, the yellow component of colour can be quantified by shining blue light through the sample and measuring how much of the light it absorbs. Each coloured effluent investigated can thus be categorised in terms of 'the amount of colour present' and absorbances at each of the seven different wavelengths recorded.

A typical exhausted dyebath liquor, a typical dyehouse end-of-pipe general site effluent and a sample that meets exactly the NRA most stringent proposed river quality objective (RQO) are defined in Table 1 in terms of their absorbances.

Using the entirely arbitrary figure of average absorbances of each sample, the substance of the challenge is apparent. To meet the strictest proposed RQO, 99.135% of the colour from the typical general site effluent would have to be removed; for the dye liquor, the comparable figure is 99.942%. This, therefore, is a major challenge.

Table 1 Typical absorbance figures

Sample	Absorbance at different wavelengths (nm) in 1 cm cell							Average absorbance
	400	450	500	550	600	650	700	
Typical dye liquor	31.7	26.04	31.47	33.49	8.48	1.62	0.008	18.983
Typical site effluent	1.423	1.529	1.965	1.715	1.514	0.772	0.62	1.283
RQO	0.025	0.015	0.012	0.010	0.008	0.005	0.003	0.0111

As has been stated, the NRA absorbance figures quoted represent the most stringent RQO so far proposed on a particular river. This is very much a worst case scenario. Actual consent figures imposed upon a discharger of textile effluent depend on several factors, especially if the discharge is to sewer and via the local sewage treatment works. In these cases dilution factors may be taken into account and so the consent figures could be somewhat more relaxed. For the purpose of analysing this particular technique, however, it is clearly appropriate to compare the capabilities with this most stringent standard.

Having established the extent of the theoretical task, it is now appropriate to consider the practical implications of attempting colour removal. A full-scale colour removal process on a given site may not only have to remove over 99% of the colour in a certain sample, but maintain this level of colour removal on sometimes large volumes and on rapidly changing effluents. The effluent or liquor to be decolorised changes in terms of colour shade, intensity, sometimes dye type and certainly treatability, as is well known; it is frequently the biggest problem in textile effluent treatment. The management of the colour-containing effluents therefore becomes a vital component in the overall strategy for a dyehouse considering dye removal.

The basic principles of this colour removal technique, involving the use of synthetic organic coagulants to chemically precipitate residual dye, can be applied either to relatively small volumes of concentrated exhausted dyebath liquors, to the much larger volumes of general site effluent (which contain not only spent dye but also all washing or scouring liquors) or even to sewage containing dye residues at a local sewage works.

Each of these requires a consistently high level of colour removal, and the aforementioned practical 'problems' facing all such techniques frequently guides the prospective user towards application of treatment at a particular effluent stream or liquor location. In either case, the above problems of effluent variability frequently arise. This paper will look at the question of how the addressing of these potential problems can assist successful full-scale colour removal from effluent streams.

3 Colour removal mechanism

The mechanism by which synthetic organic polymer coagulants remove dissolved dye residuals from liquors or effluents is best described in terms of the electrostatic attraction between the oppositely charged soluble dye and polymer molecules. Many of the most problematic dye types, such as reactive dyes, carry a residual negative charge in their hydrolysed dissolved form, and so positively charged groups on the polymers provide the necessary counter for the interaction and subsequent precipitation to occur. The result of this coprecipitation is the almost instantaneous production of very small coloured particles, which slowly grow to perhaps 2 mm in diameter. These particles, once grown, have little strength and break down if they experience any significant disturbances.

For most solid/liquid separation processes, these larger particles must have sufficient strength to survive the turbulence in such environments. This 'floc strength' can be given to precipitate agglomerations by using appropriate high-r.m.m. polyelectrolyte flocculants, which produce stable 'flocs' to perhaps 4 mm in diameter. Figure 1 shows a representation of this process.

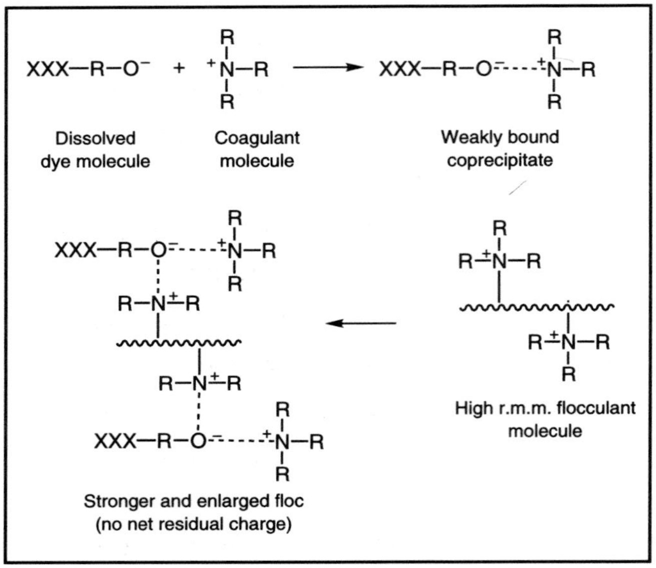

Figure 1 *Coagulation and flocculation mechanism*

3.1 Colour removal testing and results

For tests on a liquor or effluent from a particular site to have relevance on prospective colour removal at that site, a good deal of careful planning is normally required for the collection of an appropriate sample.

Where the individual main components of an overall site effluent can be obtained, this is often the preferred route providing accurate final composition ratios are available. In such cases, composite samples of each component can be collected over perhaps several days, allowing recombination during the evaluation process. This would produce samples of well-balanced liquors of each type so that views can be established on the preferred treatment location. If individual component sampling is not possible, composite samples of effluents or liquors can be collected over a suitable period of time for appropriate locations in the dyehouse effluent system.

For prospective colour removal treatment at the dyehouse, interest focuses on the end-of-pipe general site effluent and the combined dyebath liquor/rinse liquor streams (i.e. the coloured components only). Clearly, if it is not possible to access individual streams such as dye waste, rinse liquors, scour liquors or the alternative route of combined coloured components, it may be that the infrastructure at the dyehouse will accept treatment at end-of-pipe as the only option, and so efforts can be directed accordingly from the start. If treatment at different points of the system is a realistic option, then testing carried out using this procedure can frequently help to identify preferred strategy.

Laboratory tests, as recorded in the case studies in this paper, are carried out using 'jar testing' apparatus, involving 500 ml sample aliquots in typically four or six jar variable-speed multistirrer laboratory flocculators. The treatment conditions in terms of pH and mixing energy are optimised and the solutions of polymer coagulant added over a dose range to determine an optimum dose rate. Where necessary, very small dose levels of high-r.m.m. flocculant are added following a suitable contact time, to strengthen the flocs as described above.

After a period of time to allow the flocs to separate from the liquor samples, the clear liquors are extracted for analysis, notably for COD and residual colour measurement via a suitable 1 cm cell UV/VIS spectrophotometer using the NRA-specified procedure of prefiltering through a 0.45 μm membrane.

3.2 Discussion of results

The summaries of results shown in the four case studies given here have been selected and structured to highlight how a thorough sampling and laboratory testing programme can help to identify a preferred technical solution for a particular site.

In case study 1, all the key constituent liquors – pad waste, rinse liquors and the noncoloured remainder of the overall effluent – could be sampled individually, and the location of colour removal was unspecified at the time of the enquiry. The results suggested that in terms of optimum colour removal from the two selected possible locations for

Case study 1 Cotton cold pad–batch dyer using reactive dyes

Liquor constituents (see section 3.2)
The samples below reflect different stages in the process using recombined composite sampled components.

Component analysis

	Absorbance at different wavelengths (nm)							
Sample	400	450	500	550	600	650	700	Average absorbance
Pad liquor	107.1	78.9	90.4	130.4	165.8	92.7	11.3	96.7
Pad liquor + rinse	15.88	10.89	11.22	16.50	20.89	12.60	1.44	12.77
End-of-pipe	0.95	0.702	0.862	1.246	1.463	0.839	0.075	0.877

Evaluation of pad liquor plus rinses and end-of-pipe effluent

Coagulant	Dose /mg l⁻¹	Contact time/min	Appearance	No
Pad liquor + rinses	–	–	Strong deep blue	1
Magnafloc 1897	3000	10	Blue coloration	2
Magnafloc 1897	5000	10	Clear and colourless	3
End-of-pipe	–	–	Deep blue	4
Magnafloc 1897	600	10	Blue tinge	5
Magnafloc 1897	800	10	Clear and colourless	6

Results

No	COD /mg l⁻¹	Absorbance at wavelength shown (nm) in 1 cm cell							Average absorbance	Removal /%
		400	450	500	550	600	650	700		
1	5057	15.88	10.89	11.22	16.50	20.89	12.60	1.44	12.77	–
2	1850	3.04	2.105	1.576	2.017	2.437	1.494	0.136	1.829	85.68
3	1433	0.033	0.017	0.015	0.010	0.002	0.000	0.000	0.011	99.91
RQO	–	0.025	0.015	0.012	0.010	0.008	0.005	0.003	0.011	–
4	1778	0.95	0.702	0.862	1.246	1.463	0.839	0.075	0.877	–
5	708	0.021	0.016	0.019	0.011	0.004	0.002	0.000	0.010	98.9
6	694	0.011	0.008	0.007	0.004	0.001	0.001	0.000	0.005	99.5

treatment, the end-of-pipe option may well be the favoured one. The level of colour remaining after treatment of the end-of-pipe sample was well below even the 'worst case' RQO levels (p. 172), whereas that following treatment of the dye-plus-rinses liquor was only

Case study 2 Multifabric dyer using mainly disperse dyes via continuous padding process

Liquor constituents (see section 3.2)
Samples below reflect different stages in the process with as far as possible equivalent dye type using composite samples.

Component analysis

| Sample | Absorbance at different wavelengths (nm) in 1 cm cell | | | | | | | Average absorbance |
	400	450	500	550	600	650	700	
Dye liquor	93.3	81.2	104.7	149.2	155.7	89.8	47.6	103.07
Dye liquor + rinses	6.15	5.15	6.91	10.51	11.8	6.01	2.82	49.35
End-of-pipe	0.619	0.417	0.294	0.218	0.173	0.125	0.083	0.276

Evaluation of dye liquor plus rinses and end-of-pipe effluent

Coagulant	Dose /mg l^{-1}	Contact time/min	Appearance	No
Dye liquor + rinses	–	–	Dark blue	1
Magnafloc 1897	1500	10	Slight yellow tinge	2
Magnafloc 1897	2000	10	Clear and colourless	3
End-of-pipe	–	–	Murky green	4
Magnafloc 1897	300	10	Brown residual	5
Magnafloc 1897	400	10	Pale straw residual	6

Results

| No | COD /mg l^{-1} | Absorbance at wavelength shown (nm) in 1 cm cell | | | | | | | Average absorbance | Removal /% |
		400	450	500	550	600	650	700		
1	7800	6.15	5.15	6.91	10.51	11.8	6.01	2.82	49.35	–
2	1670	0.03	0.007	0.000	0.000	0.000	0.000	0.000	0.005	99.99
3	1588	0.017	0.004	0.000	0.000	0.000	0.000	0.000	0.003	99.99
RQO	–	0.025	0.015	0.012	0.010	0.008	0.005	0.003	0.011	–
4	1968	0.619	0.417	0.294	0.218	0.173	0.125	0.083	0.276	–
5	1619	0.155	0.052	0.029	0.015	0.012	0.008	0.005	0.039	85.87
6	1558	0.112	0.016	0.008	0.004	0.003	0.003	0.003	0.021	92.39

comparable with the RQO level. COD removal was comparable in the two cases, but in terms of the amount of coagulant required to achieve optimum colour removal, on an equivalent pro rata basis, the end-of-pipe option could be calculated to require less.

Case study 3 Multifabric dyeing/printing using reactive dyes and pigments

Liquor constituents (see section 3.2)
Samples below reflect different stages in the process with as far as possible equivalent dye type using composite samples.

Component analysis

Sample	Absorbance at different wavelengths (nm) in 1 cm cell							Average absorbance
	400	450	500	550	600	650	700	
Dye liquor + rinses	1.692	1.574	1.879	1.718	1.133	0.505	0.313	1.250
Noncolorants	0.072	0.036	0.024	0.017	0.011	0.009	0.007	0.025
End-of-pipe	0.941	0.827	1.147	1.132	0.548	0.355	0.137	0.727

Evaluation of colour kitchen sample and also end-of-pipe effluent

Coagulant	Dose /mg l^{-1}	Contact time/min	Appearance	No
Dye liquor + rinses	–	–	Deep murky purple	1
Magnafloc 1897	800	10	Very pale pink	2
Magnafloc 1897	1000	10	Clear and colourless	3
End-of-pipe	–	–	Deep red	4
Magnafloc 1897	600	10	Very pale pink	5
Magnafloc 1897	800	10	Clear and colourless	6

Results

No	COD /mg l^{-1}	Absorbance at wavelength shown (nm) in 1 cm cell							Average absorbance	Removal /%
		400	450	500	550	600	650	700		
1	6350	1.692	1.574	1.879	1.718	1.133	0.505	0.313	1.250	–
2	3265	0.108	0.091	0.080	0.082	0.061	0.057	0.048	0.075	94.0
3	2949	0.004	0.010	0.008	0.005	0.006	0.001	0.002	0.005	99.60
RQO	–	0.025	0.015	0.012	0.010	0.008	0.005	0.003	0.011	–
4	3005	0.941	0.827	1.147	1.132	0.548	0.355	0.137	0.727	–
5	1215	0.042	0.027	0.023	0.023	0.017	0.013	0.006	0.022	97.0
6	1105	0.015	0.001	0.002	0.003	0.000	0.001	0.002	0.003	99.53

Case study 2 is a similar example, this time using predominantly disperse dyes. At this site individual components of the overall site effluent could not be obtained, but combined dye plus rinses, and obviously end-of-pipe effluent, were available. In this case, treating the

Case study 4 Settled primary sewage containing reactive dyes

Liquor constituents (see section 3.2)
Samples taken from a large sewage treatment works with consistent, but usually quite low levels of colour from local textile industry.

Evaluation of sewage sample

Coagulant	Dose /mg l⁻¹	Contact time/min	Appearance	No
Untreated	–	–	Dirty brown tinge	1
Magnafloc 368	5	10	Clear and colourless	2
Magnafloc 368	10	10	Clear and colourless	3

Results

No	COD /mg l⁻¹	Absorbance at wavelength shown (nm) in 1 cm cell							Average absorbance	Removal /%
		400	450	500	550	600	650	700		
1	44	0.032	0.022	0.020	0.022	0.022	0.009	0.001	0.018	–
2	44	0.016	0.008	0.007	0.006	0.005	0.002	0.001	0.006	65.03
3	44	0.013	0.006	0.004	0.003	0.002	0.001	0.001	0.004	76.50
RQO	–	0.025	0.015	0.012	0.010	0.008	0.005	0.003	0.011	39.34

coloured liquors only may be the preferred option since on subsequent combination with the noncoloured liquors, analysis of the end-of-pipe effluent showed consistently a difficulty in removing the yellowness indicated by the 400 and 450 nm absorbances. COD removal was higher on the coloured liquors and the amounts of coagulant required on an equivalent pro rata basis were similar (but for a poor result on the end-of-pipe liquor).

Case study 3 resembles case 2 in terms of the sampling options available, but this time the liquor contained a mixture of reactive dye and pigment print. Although the analysis shows both effluents could be treated satisfactorily, the end-of-pipe option is the preferred route in this case because of the site's desire to treat all its effluent in order to keep trade effluent charges to a minimum. In both this case study and case study 1 treatment performance was exceptionally good on the end-of-pipe effluent; in case 2 it was equally good on the coloured-only stream. Clearly, if these levels of performance can be maintained consistently (and this is discussed in the next section), then the COD reductions alone should offer genuine savings on trade effluent discharge rates, and the very high levels of colour removal not only allow consent passes, but also create openings for limited water re-use in selected areas.

Case study 4 shows how the colour removal technique can be applied to sewage containing dyes, and although the amount of coagulant required varies from one sewage treatment works to another, this case showed good results with very low doses. Various factors may make colour removal at the sewage treatment works the preferred or perhaps the only option. Many dyehouses simply do not have the space or the infrastructure to allow effluents to be routed to a single location for treatment before discharge. In other situations, a certain number of dyehouses may all discharge to the same sewage treatment works; if each treats on its own site, that same number of capital schemes will be required. Clearly, colour removal at the sewage treatment works would render those combined capital expenditures unnecessary, being replaced by a single chemical dosing system at the sewage treatment works. The counter-arguments lie in the limitations to an individual dyer's control over his future trade effluent charges and lack of the possible water re-use option and also, more importantly, the corporate policy of the relevant water services company over such issues.

Having established the technical aspect of the treatability of carefully sampled effluents and liquors, section 4 of this paper outlines how this laboratory-scale success might be applied in practice in full-scale plant.

4 Colour removal process plant

Effective process plant in which to carry out dye removal consists of two basic components, regardless of the liquor or effluent types to be decolorised:
— effluent or liquor collection and balancing
— colour removal from balanced effluent or liquor.

4.1 Effluent/liquor collection and balancing

This step is possibly the most important stage in the whole process of colour removal in full scale. Successful laboratory evaluations using the recombined composite split component technique (where possible) are the best method of indicating the potential suitability of a site to polymer coagulation as a method of colour removal to desired standards. In actually designing a full-scale plant to achieve the same in practice, the key to success may well be to use this same method of feed liquor preparation on site in order to stabilise the overall effluent or liquor to be treated to the greatest degree of consistency. Often, such segregation of streams prior to balanced blending may be impossible: the site may simply have a wildly variable single effluent stream. In such cases large-scale balancing tanks representing perhaps 24 h of the flow may be required to provide a reasonably consistent feed to the colour removal plant. Alternatively, if the variations are too great, the argument may be swayed in favour of processing the dye-containing

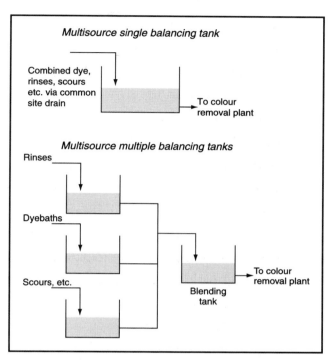

Figure 2 *Effluent/liquor collection and balancing*

liquors only. Thus the design of the full-scale effluent/liquor collecting and balancing process plant can be very site-specific. Figure 2 covers some of the basic options.

4.2 Colour removal plant

Once the equalising or balancing of the effluent feed has been successfully carried out, the capabilities of the colour removal plant are greatly improved. The key steps in the colour removal process are:

- dye precipitation
- precipitate particle agglomeration
- solid–liquid separation.

In the dye-precipitation stage, the balanced liquor or effluent is received, the pH is optimised and the correct dose of coagulant is added; precipitation then takes place within a stirred chamber or similar. When the required efficiencies of colour removal exceed 99%, highly accurate dosing is required to maintain performance. Here lies the real benefit of using well-balanced effluent or liquor feeds: once the dose levels have been applied with a high degree of precision, continuous adjustments will not be required (providing the effluent composition is relatively constant).

Once the precipitate has formed, the next stage is to strengthen and enlarge these particles to allow subsequent separation. This is normally a simple process requiring a very low dose of high-r.m.m. organic flocculant. The larger flocculated particles settle much more easily than the initial precipitate; more usually, they are induced to float by dissolved air bubbles carrying the flocs upwards (the dissolved air flotation process).

Once these larger flocs have separated from the clear liquor, the extra strength imparted by the flocculant allows the resulting sludge to be removed by 'scraping' – in the case of dissolved air flotation, without re-dispersal of the particles.

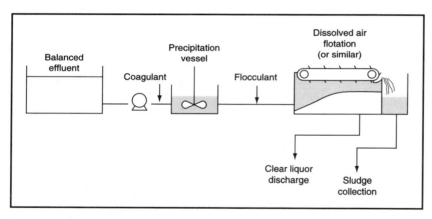

Figure 3 *Colour removal plant*

Figure 3 illustrates such a colour removal process and details the key components.

To sum up, the colour removal process plant required for on-site treatment is essentially the same whether 'end-of-pipe' general site effluents are treated, or the smaller volumes of coloured components only. Typically only the sizes of the individual components of the plant vary, according to those differences in flow rates treated.

5 Cost and payback parameters

The costs of a colour removal process may be broken down as follows:

- civil costs: infrastructure changes and new site preparation
- equipment costs: purchase and installation of plant
- running costs: chemical, power, water and sludge disposal.

Equally the payback parameters are several in origin, but start with the need for effluent treatment in certain cases to meet legal and regulatory requirements. They may extend to:

- trade effluent costs: reductions in COD and suspended solids (colour is becoming a charge factor also)
- water re-use: reductions in clean water costs if on-site uses can be found for treated effluent.

The costs associated with civil work and the purchase of equipment are extremely difficult to estimate. There is a tendency for each case to be unique, certainly from the point of view of civil preparation work. The colour removal plant alone may cost between £100 000 and £200 000 for a medium or large user of end-of-pipe treatment, but the cost of the effluent or liquor collection and balancing system varies dramatically from site to site.

As far as running costs are concerned, power and water requirements will again vary from one process to another, but chemical costs normally fall into bands associated with each treatment option (Table 2). In each case, the bulk of the cost is attributable to the main colour-removing coagulant. The cost of flocculation and pH adjustment is normally a quite small proportion of these figures.

Table 2 Chemical costs

Treatment option	Typical cost band/£ m^{-3}
Coloured liquors only	0.80–2.00
End of pipe	0.30–0.90

Sludge disposal is the other major on-going expenditure. The plant described in this paper would typically produce a sludge of approximately 5% dry solids. Such a plant processing, say, 1000 m^3 per day of end-of-pipe effluent might produce 10 m^3 of this 5% sludge daily. Removing and disposing of this material would cost between £10 and £15 m^{-3} to be removed and disposed of, by a variety of routes, in its basic form; the cost would be significantly less on a pro rata basis if the sludge can be further dewatered to a cake on site.

The payback possibilities are somewhat easier to calculate given an expected treated COD and suspended solids level and can simply be calculated using the appropriate trade effluent charge calculation. Such a calculation might be as shown in Eqn 1:

$$\text{Charge } (£ \text{ m}^{-3}) = R + V + B\frac{O_t}{O_s} + S\frac{S_t}{S_s} \tag{1}$$

where R = conveyance charge

V = volumetric charge

B = COD factor

S = suspended solids factor

O_t = average COD

O_s = specified standard COD

S_t = average suspended solids

S_s = specified standard suspended solids.

Assuming that R = £0.12, V = £0.11, B = £0.13 and S = £0.07 (all per m³) and O_t = 300, O_s = 400, S_t = 500 and S_s = 400 (all mg l⁻¹), then the overall charge is £1.293 m⁻³.

Suppose, however, that after treatment average COD is reduced by 50% to 1500 mg l⁻¹ and average suspended solids is reduced to 100 ppm; the new overall charge would then be £0.735 m⁻³. This hypothetical example would therefore create a payback possibility of £0.558 m⁻³ with which to offset the additional cost of effluent treatment. Also, the cost of buying clean water for process use in the first place may cost a site in excess of £0.50 m⁻³, so if any of the treated effluent can be used to replace some of the bought-in clean water further savings are clearly possible.

6 Conclusions

This paper has described one of several techniques for the removal of colour from dyehouse effluent. The precipitation of dye using synthetic organic coagulants in this way is already used by dyehouses in the United Kingdom, in Europe and elsewhere to good effect, with increasing interest from other sites wishing to protect themselves against legislation and the growing costs of discharging trade effluent.

The efficiency of colour removal using this technique is increasing. Investment in research and development is bringing new products and procedures into the market-place, and the case studies reported show just how powerful the technique now is.

CHAPTER FOURTEEN

Effluent treatment using chemical flocculation

Brian W Iddles

1 Effluent treatment in the textile industry

Several plants have been developed for the removal of both reactive and disperse dyes, operating in the UK, Ireland and the USA. Other work has been carried out on COD and pesticide removal from scouring and desizing operations.

The system is based upon treating the raw effluent with a measured amount of either a single or a mixed dose of inorganic chemical. This chemical mixes and then dissolves in the effluent. A second chemical is then added to increase the pH of the solution, causing precipitation of both the primary treatment chemical and the offending pollutant. A further pH adjustment may then be necessary to lower the pH sufficiently to enable discharge to the public foul water sewer or watercourse.

The effectiveness of the treatment package for some specific effluents are summarised in Table 1.

1.1 Clarification using a tangential flow separator

Techspan Associates together with the Clean Water Co. Ltd has been involved in the development of water purification plants that use very little energy and occupy small amounts of space. This is achieved by utilising simple principles which are well proven and patented. The separation system originally designed was intended for use in the third world where technology is scarce or non-existent. The simplicity of the design has been retained in a large number of plants now in use for various applications in several industries.

185

Table 1 Some typical effluents in different textile industry sectors

Wool scouring effluent/mg l^{-1}

	Inlet	Outlet[a]	Reduction/%
COD	27 000	7 650	67
COD	1 328	195	85
Suspended solids	4 240	142	96
Suspended solids	1 640	33	98

Carpet manufacturing effluent/mg l^{-1}

	Raw	Treated[b]
Permethrin 1	63	0.6
Permethrin 2	192	1.4
Cyfluthrin	193	4.3

Textile dyeing effluent/mg l^{-1}

Raw		Treated[c]	
COD	Colour	COD	Colour
911	Purple	374	Colourless
687	Yellow	344	Colourless
705	Red	231	Colourless
566	Purple	391	Colourless
893	Blue	406	Colourless
411	Green	369	Colourless

a Chemical treatment cost = £0.15 m^{-3}
b Because the resultant effluent was free from suspended solids and other contaminants, it could then be passed through a carbon filter to remove the traces of permethrin present without exhausting the carbon
c A secondary product of the colour removal is the reduction in COD; a reduction of at least 50% is normally achieved, representing a large saving in effluent disposal costs (see text)

The systems are based on the Cleanwater tangential flow separator, which is a highly efficient sedimentation/clarification unit (Figure 1). In this unit liquids to be treated are introduced tangentially into a cylindrical vessel fitted with a cone bottom. The vessel is designed to control the flow of liquid around the outer wall of the vessel in a spiral flow. The complex vortex flow patterns induced into the flow by the special lid and head frame design cause suspended particles to rotate on their own axis, optimising the effect of gravity, so that solid matter rapidly passes to the bottom of the vessel to be continuously discharged as a slurry into a sludge thickening tank. Clarified liquid leaves the system from tne centre of the top of the separator, and any floating matter is held behind a collar.

The system works efficiently under a hydraulic head of 1 m or less. The separator is not a hydrocyclone or a centrifuge. Compared with a traditional sedimentation unit, the tangential flow separator performs better than a traditional plant whilst occupying as little as 20% of the area (Table 2). Normal operation results in 90% of the liquid leaving the unit as cleaned flow, with the remaining 10% continuously discharging to the sludge thickener. After passing through the thickener the supernatant is returned to the inlet of the system, and sludge of around 5% solids is obtained.

Due to the increasing cost of sludge disposal further plant is normally installed to dewater the sludge to around 30% solids content. This plant may vary from a simple manual filter press to an automated centrifuge, depending on the nature of the sludge and the customer's requirements. Because the system operates from a 1 m head the power consumption is very little compared with other treatment systems; Table 3 lists actual figures for an installation processing 70 m³ h⁻¹ on a continuous basis. As there are few moving parts, the maintenance requirement is also very low, typically two man hours per day including chemical make-up and filter pressing.

Trials were carried out using a pilot plant alongside an existing DAF unit. The effectiveness and operating costs are compared in Table 4; although in this instance food-processing effluent was being treated, a comparison using textile effluent could be expected to lead to similar results.

2 Practical applications

There is increasing pressure on the textile industry to install effluent treatment systems, but because of the high capital costs involved some companies view the prospect with alarm. Indeed it has been widely reported that many old-established companies face closure because of the prohibitive cost of effluent treatment.

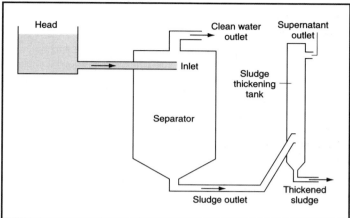

Figure 1 *Tangential flow separator: schematic diagram*

Table 2 Comparison between traditional sedimentation tank and tangential flow separator (TFS)

	Surface loading /l s⁻¹ m⁻¹	Suspended solids removal/%	COD reduction/%
Sedimentation tank	0.15	51	40
TFS without coagulant	0.8	62	42
TFS with coagulant	0.8	91	74

Table 3 Power consumption in tangential flow separator

	Power consumption/kW h⁻¹				
Effluent flow rate /m³ h⁻¹	Lift pump	Sludge transfer pump	Chemical dosing pump	Control panel	Total
7	0.75	0.55	0.24	0.2	1.74
15	1.1	0.55	0.24	0.2	2.09
45	3.5	0.75	0.36	0.3	4.91
85	6.5	1.1	0.36	0.4	8.36

Table 4 Effluent treatment: comparison of DAF and Techspan installations

	Inlet to plant	DAF plant	Techspan treatment
COD	2 514	454	166
	4 860	660	232
	8 740	825	450
	12 500	1 120	660
Yearly operating costs[a]		£5 750	£330.00
Chemical usage[b]			£250
Annual savings[c]			£18 420

a Power + labour
b Weekly savings
c Compared with DAF unit

Techspan treatment systems are not only designed with relatively low capital and running costs but, by virtue of the reduction in suspended solids and COD, a payback time on the plant can often be calculated. Consider a hypothetical example:

Average COD of raw effluent	=	900
Average COD of treated effluent	=	450
Yorkshire Water regional strength	=	965
Yorkshire Water biological charge	=	
		£0.1792 m^{-3}

Charge for raw effluent	=	$900/965 \times 0.1792$
	=	£0.1671 m^{-3}
and		
Charge for treated effluent	=	$450/965 \times 0.1792$
	=	£0.0849 m^{-3}
Saving	=	£0.0822 m^{-3}

i.e. on a continuous discharge of 50 m^3 h^{-1}, six days a week, 48 weeks a year, the saving is £28 400.

2.1 Water recycling

Before considering the direct financial advantages of partial recycling of treated effluent, some mention should be made of BS 7750 and eco-labelling. Most textile dyeing and finishing in the UK is by commission, and therefore the finisher concerned may mistakenly believe that eco-labelling has no part in his operation. End-users, however, are already beginning to ask questions on the environmental performance of their suppliers and the use of recycled water gives a 'low score', and therefore a competitive edge, on any environmental questionnaire.

It is accepted that a large number of dyers and finishers have their own private water supply, normally a borehole, and this means a relatively cheap source of water, but there are still large savings to made on water recycling. If partial recirculation were introduced (a 40% figure has been used) on the above example of 50 m^3 h^{-1}, the saving on effluent disposal charges would be [0.1671 − (0.0822 × 60/100)] = £0.1178 m^{-3}.

On this basis, and assuming a charge of £0.50 m^{-3} for water supply charges, the actual saving on water charges is [(0.50 × 60/100) + 0.1178] = £0.4178 m^{-3}, i.e. on a continuous discharge of 50 m^3 h^{-1}, six days a week, 48 weeks a year, the saving is £144 390.

2.2 Tertiary treatment

On occasions there is a requirement to reduce the COD to a level lower than that achievable by chemical flocculation, usually where the effluent produced is discharged directly to a watercourse. The problem is easily overcome by using a patented process called a biofilter which uses a biofixative, Procede COR (Figure 2).

The biofilter itself is a cylindrical tank; typically this is 2 m in diameter and 2.5 m high for a flow of 15 m^3 h^{-1}, and packed with the biofixative. The chemically treated effluent is introduced at the bottom of the tank and discharges leave from the top. A contact time of less than 1 h normally gives an outlet COD of less than 100 mg l^{-1}.

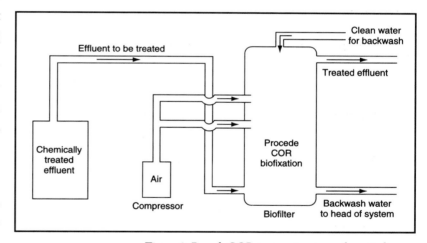

Figure 2 *Procede COR treatment system: schematic diagram*

3 Procede COR biofixation

The process of biofixation, the fixing of bacteria and fungi to mineral supports by chemi-osmoregulation, represents a considerable advance in the application of biotechnology.

The effectiveness of bioaugmented treatments is subject to several physico-chemical limiting constraints affecting the reactivation and survival of micro-organisms in the toxic wastes they are to degrade. Many of these problems are overcome by the process of biofixation. It depends upon fixing selected strains of natural micro-organisms to carefully chosen mineral supports by chemi-osmoregulation. The combination of bacteria and support gives the micro-organisms superior capabilities of activity, toxin resistance and reproduction, substantially greater than either free-living or bioaugmented organisms.

3.1 Principles of action

Broadly speaking, two types of action take place, both providing powerful synergistic effects: physico-chemical and biological.

Physico-chemical action

(a) The ionic effects of the mineral support lead to coagulation of organic pollutants around the biofixation particles.
(b) Ammonium ions, sulphides and mercaptans are adsorbed.
(c) The system acts as a pH buffer.
(d) Ion exchange regulates the production of undesirable concentrations.
(e) The potassium aluminosilicates of the support block certain harmful metallic elements, by exchange with potassium ions.
(f) Certain elements are precipitated from solution.

Biological action

The bacteria fixed to the mineral support have been selected because of their ability to destroy certain pollutants.

(a) Chemical micro-pollutants are degraded, including some that resist degradation by the bacteria in natural water treatment plants.
(b) Because of the nature of the support, the bacteria have improved toleration of certain toxins, including chlorine.

(c) The speed of reproduction and enzyme production is greater for biofixed bacteria than for 'free-floating' bioaugmented bacteria.

(d) Biofixed bacteria will out-compete obstructive (e.g. filamentous) bacteria for nutrients.

(e) The support becomes a base around which a floc can grow.

(f) Sludge quality is improved – there is a powerful degrading action on organic matter.

(g) Levels of activity at lower oxygen levels are enhanced, thus making possible considerable energy savings.

4 Example 1 – commission dyer

Company 1 is a large commission dyer. It has an average effluent flow of 70 m^3 h^{-1} and as there is no foul sewer available this is discharged to river. Both effluent and sewage produced on site are treated with an activated sludge plant with dissolved air flotation. The effluent continually failed to meet its colour limits and the company was prosecuted on several occasions by the NRA.

After laboratory and pilot-plant trials a Cleanwater tangential flow separator with chemical pre-treatment was installed to deal with the biologically treated effluent. The effluent only once slightly exceeded its consent limit (Table 5). As the water used on site was from the company's own borehole and the effluent was discharged to a watercourse, there was no financial advantage in water recycling.

Table 5 Final effluent: compliance with NRA colour consent

	Absorbance at different wavelengths/nm						
	400	450	500	550	600	650	700
January	0.019	0.015	0.010	0.004	0.003	0.002	0.000
February	0.017	0.015	0.011	0.005	0.003	0.002	0.000
March	0.027	0.026	0.023	0.014	0.009	0.008	0.005
April	0.021	0.020	0.019	0.007	0.005	0.005	0.003
May	0.014	0.008	0.007	0.006	0.005	0.005	0.003
June	0.057	0.044	0.035	0.022	0.012	0.008	0.005
NRA consent limit	0.060	0.040	0.035	0.025	0.025	0.015	0.010

After one year's operation the plant configuration was changed and the tangential flow separator was used before the biological plant. The separator continued to remove not only colour but also a high percentage of COD and suspended solids. By reducing the organic loading on the biological plant the company were able to put a much larger flow through the system, and were therefore able to increase production by 30% without adverse effect on their final effluent.

5 Example 2 – denim washer

Company 2 is a large denim washer, dyer and finisher in the north of England. It discharges an average of 80 m^3 h^{-1} into the public foul water sewer, and the company is subject to a consent issued by Yorkshire Water. It has no private water supply, all its water coming from Yorkshire Water mains supply. Its annual expenditure on water supply and effluent disposal exceeds £5 000 000. As well as anticipating a colour limit being imposed on it, the company was interested in water recycling.

Techspan Associates was commissioned to produce a treatment system that would decolorise the effluent. This would enable the company to comply with any colour consent issued by Yorkshire Water and also allow 50% of this treated effluent to be recycled.

The system designed, now operating, treats the effluent to a standard that complies with the present Yorkshire Water consent (i.e. no limit to colour, just COD and suspended solids) and also removes sufficient colour to enable recycling to take place. This represents a treatment cost saving over a colourless effluent of around £0.12 m^{-3}. When Yorkshire Water finally decides on the colour limit that will be imposed this can be easily achieved by adjusting the chemical dosage.

The portion of treated effluent that is recycled passes first through a sand filter to give the effluent a final polish and then to a tank, where it is blended with mains water to re-use in the process. The total cost of the package including sand filter, filter press and all installation to treat a flow of 80 m^3 h^{-1} was less than £250 000. Allowing for chemical usage, sludge disposal and maintenance costs, the payback time is less than two years.

Part Four Novel technologies

Chapter 15

Dyeing in nonaqueous systems, *David M Lewis*

Various solvent-dyeing systems have been evaluated and rejected by the dye application industry, mainly because of problems of air pollution. In 1991 a 'safe' solvent-dyeing system was launched, based on the use of supercritical carbon dioxide as a solvent for disperse dyes. Progress to date is critically reviewed, and future directions assessed.

Dyeing in nonaqueous systems

David M Lewis

1 Introduction

The problem of water pollution by the dyeing industry has intensified research effort directed at using nonaqueous media. Twenty or thirty years ago there was much concentrated activity in the use of organic solvents to replace water as a dye application medium; in particular, perchloroethylene and trichloroethylene seemed to be the solvents of choice [1,2]. This approach was seen to be flawed, however, since it merely transformed the water pollution problem into one of air pollution. Recovery of organic solvent can never be 100% efficient, and the solvent residues are lost to the atmosphere. In particular, perchloroethylene vapour is so dense that it will remain near the ground until photodegraded; one of its decomposition products is phosgene. Other organic solvents are now viewed with suspicion too, since they may contribute to urban smog and greenhouse gases.

A further drawback to the various processes for organic solvent dyeing was that most were designed for the dyeing of polyester fibres with disperse dyes; since the major textile fibre is cotton (some 52% of the textile fibre market), clearly the development of a nonaqueous process for dyeing this fibre should be a priority. This is especially true since most problems of coloured effluent arise from the dyeing of cotton with reactive dyes, because of the relatively poor fixation achieved in deep shades [3]. Solvent dyeing processes for the polyamides, wool and nylon were also investigated, using mixtures of perchloroethylene with booster solvents such as methanol, acetic acid and even water. None of these systems was seen as viable in practice, however.

2 Dyeing from supercritical fluids

At ITMA 1991, interest in solvent dyeing was revived when Joseph Jasper, from Velen, demonstrated a sample machine for dyeing polyester yarn in supercritical carbon dioxide. This machine had been built as a result of initial research at the Deutsches Textilforschungs-Zentrum Nord-West e.v., Krefeld [4,5]. Subsequently Ciba became involved in the provision of suitable dyes [6], fluorescent brightening agents and other auxiliaries; Joseph Jasper cooperated and produced suitable machinery. Before describing the dyeing process it is necessary to briefly discuss the properties of supercritical fluids.

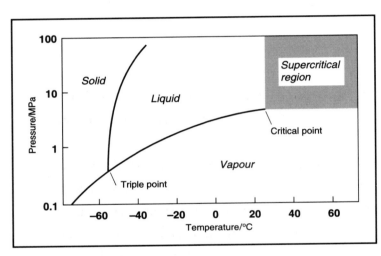

Figure 1 *Phase diagram of carbon dioxide*

How supercritical fluids are produced by the effects on gases and liquids changes in pressure and temperature is illustrated in Figure 1, the phase diagram of carbon dioxide. The curves in this figure represent interfaces between phases; at the triple point (ca. – 60 °C and 0.5 MPa pressure), all three phases may coexist. Above the triple point, an increase in temperature drives liquid into the vapour phase whilst an increase in pressure drives vapour back into liquid. This situation continues until the critical point is reached; above the temperature at this point, the *critical temperature*, the gas cannot be liquefied however high the applied pressure. In the case of carbon dioxide, the critical point is at 31 °C and a pressure of 7.3 MPa.

Above its critical temperature, a gas retains the free mobility of the gaseous state but with increasing pressure its density will increase towards that of a liquid. Such a highly compressed gas is termed a *supercritical fluid*, and as such it combines the valuable properties of both liquid and gas. Its solvating power is proportional to its density, whilst its viscosity remains comparable to that of a normal gas; the 'fluid' thus has remarkable penetration properties. The increase in solvating power with increasing density is crucial to the dyeing process; it has an enormous effect on the dissolution of disperse dyes in supercritical carbon dioxide.

The machinery currently available from Joseph Jasper is for dyeing polyester yarn in package form. A typical dyeing procedure may be summarised:

- set up with goods and a charge of pure disperse dye
- run in supercritical carbon dioxide (at – 5 °C)
- raise temperature and pressure to 130 °C and 30 MPa (to achieve solubilisation of disperse dye)
- gradually reduce pressure in order to reduce the solubility of the disperse dye in the supercritical fluid
- recover carbon dioxide.

In this procedure, which takes about 30–45 minutes, the levelness of dyeing is controlled by the pressure reduction programme. Virtually 100% dye uptake may be obtained and the need for reduction clearing is eliminated.

Currently, it is claimed that the system is suitable for dyeing polyester, aramid and polypropylene fibres, special dialkylaminoanthraquinone dyes being necessary for the last-named.

It is doubtful whether such technology can make a full breakthrough unless it could be applied to cellulosic fibres, and Saus *et al.* evaluated different carbodiimide and resin pretreatments designed to make the cellulosic fibre more hydrophobic and hence more readily dyed with disperse dyes in the supercritical carbon dioxide system [7]. Clearly there are possibilities of development in this area, but a feasible system will probably require cellulose modification and dyeing with disperse dyes simultaneously, or at least sequentially in the same system.

3 References

1. K Gebert, *J.S.D.C.*, **87** (1971) 509.
2. B Milicevic, *Text Res. J.*, **39** (1969) 677.
3. I G Laing, *Rev. Prog. Col.*, **21** (1991) 56.
4. W Saus, D Knittel and E Schollmeyer, *Text. Res. J.*, **63** (1993) 135.
5. E Schollmeyer, D Knittel, H-J Buschmann, G M Schneider and K Poulakis, German Offen. DE 3906724 A1.
6. D Werthemann, W Schlenker and U Beines, *Textil Praxis Int.*, **46** (1991) 932.
7. W Saus, S Hoger, D Knittel and E Schollmeyer, *Textilveredlung*, **28** (1993) 38.

Index